OBJECT ORIENTED PROGRAMMING USING C++

Lalit Kishore Arora
M.Tech (C.S.), MCA, B.Sc (Comp. Sc.)
Asstt. Professor, MCA Department
Ajay Kumar Garg Engineering College, Ghaziabad

Prof. (Dr.) Vikesh Kumar
M.Tech , Ph.D.
Director
Vidya College of Engineering, Meerut

S.K. KATARIA & SONS®
Publisher of Engineering Books
4885/109, Prakash Mahal, Dr. Subhash Bhargav Lane,
Opposite Delhi Medical Association, Daryaganj, New Delhi-110002 (INDIA)
Phone: +91-11-43551243; Mobile: +91-9871775858, 7678425209
e-mail: katariabook@gmail.com, katariabooks@yahoo.com
Website: www.skkatariaandsons.com
(e-books, subscriptions and bulk buying also available on our website)

Published by:

S.K. Kataria & Sons®

4885/109, Prakash Mahal, Dr. Subhash Bhargav Lane,
Opposite Delhi Medical Association, Daryaganj, New Delhi-110002 (INDIA)
Phone: +91-11-43551243; Mobile: +91-9871775858, 7678425209
e-mail: katariabook@gmail.com, katariabooks@yahoo.com
Website: www.skkatariaandsons.com

Send all correspondence to : **M/s S.K. Kataria & Sons,** New Delhi
Email: katariabook@gmail.com

First Edition: February, 2010; Second Edition: January, 2011; Reprint: 2026

Laser Typesetting by : Plus Computers, Delhi *Printed at :* New A.S. Printing Press, Delhi

PREFACE

This book Object Oriented Programming using C++ is written to serve as a textbook for B.Tech. IV Semester students of Maharishi Dayanand University, Rohtak.

Besides this specific objective this book will also be a useful textbook for BCA and MCA students of other universities.

This book also work for the programmer who wish to know more about C++ language and object oriented programming. It explains in a simple and easy-to-understand style the what, why and how of object oriented programming with C++. The book assumes that the reader is already familar with C language, although he or she need not be an expert programmer.

This book dedicated to programming in C++, which covers all features of object oriented programming such as object, classes, inheritance and polymorphism.

In each chapter, sufficient worked examples have been given. At the end of each chapter, solved problems and programs have been given.

We have also provided the solution of all Lab assignment in the last. We have tried to solve all problems defined in the assignments.

Appendix is divided into four parts. Appendix A is given for your practice in which program codes are given, you have to predict its output or find the errors. Appendix B deals with interview related questions. Appendix D provides you tips for programming in C++. And Appendix D provides names of reference books.

The book provides numerous examples, illustrations and complete programs. The sample programs are meant to be both simple and educational. Whenever necessary, pictorial descriptions of concepts are included to improve clarity and facilitate better understanding.

The author will feel highly obliged to all readers for their constructive suggestions healthy criticism of the book which will go a long way in the improvement of the text.

—Lalit Arora
—Dr. Vikesh Kumar

ACKNOWLEDGEMENT

It is a pleasure to be able to thank all the people who helped us turn the first edition of Object Oriented Programming using C++ from a list of good ideas into a real book.

We are pleased to express our thanks to the following Academician for allowing us to work under their valuable guidance:

- Dr. Vinod Kumar Sharma, Professor, Gurukul Kangri Vishwavidyalaya, Haridwar
- Dr. Karamjeet Bhatia, Professor, Gurukul Kangri Vishwavidyalaya, Haridwar
- Dr. Rajkumar Bhatia, Professor, Gurukul Kangri Vishwavidyalaya, Haridwar
- Dr. Atul Kumar Jain, Chairman, Ideal Institute of Technology, Gzb.
- Prof. K.S. Mehta, Professor, A.K.G.E.C., Ghaziabad

Our special thanks goes to Mr. Ankit Gupta, Software developer, IBM India, Gurgoan, Mr. Harsh Vardhan, Lecturer, M.Tech, IIT, Delhi, Mr. Vikas, who played a major role in scripting this book.

We are greatly intended to Mr. Sanjeev Kataria and other publishing professionals at S.K. Kataria and Sons for their active involvement throughout the production process and for bringing out the book in its present from in record time.

—Authors

Dedicated
to My Sisters
and
My Brother
for their Bright Future

—Lalit Kishore Arora

CONTENTS

C++ PROGRAMMING LAB.

Q.1. Raising a number n to a power p is the same as multiplying n by itself p times. Write a function called power() that takes a double value for n and an int value for p, and returns the result as double value. Use a default argument of 2 for p, so that if this argument is omitted, the number will be squared. Write a main() function that gets values from the user to test this function.

Q.2. A point on the two-dimensional plane can be represented by two numbers: an X coordinate and a Y coordinate. For example, (4,5) represents a point 4 units to the right of the origin along the X-axis and 5 units up the Y-axis. The sum of two points can be defined as a new point whose X coordinate is the sum of the X coordinates of the points and whose Y coordinate is the sum of their Y coordinates.

Write a program that uses a structure called point to model a point. Define three points, and have the user input values to two of them. Than set the third point equal to the sum of the other two, and display the value of the new point. Interaction with the program might look like this :

Enter coordinates for P1: 3 4

Enter coordinates for P2: 5 7

Coordinates of P1 + P2 are : 8, 11

Q.3. Create the equivalent of a four function calculator. The program should request the user to enter a number, an operator, and another number. It should then carry out the specified arithmetical operation: adding, subtracting, multiplying, or dividing the two numbers. (It should use a switch statement to select the operation). Finally it should display the result. When it finishes the calculation, the program should ask if the user wants to do another calculation. The response can be 'Y' or 'N'. Some sample interaction with the program might look like this.

Enter first number, operator, second number: 10/ 3 Answer = 3.333333

Do another (Y/ N)? Y

Enter first number, operator, second number 12 + 100

Answer = 112

Do another (Y/ N) ? N

Q.4. A phone number, such as (212) 767-8900, can be thought of as having three parts: the area code (212), the exchange (767) and the number (8900). Write a program that uses a structure to store these three parts of a phone number separately. Call the structure phone.

Create two structure variables of type phone. Initialise one, and have the user input a number for the other one. Then display both numbers. The interchange might look like this :

Enter your area code, exchange, and number: 415 555 1212

My number is (212) 767-8900

Your number is (415) 555-1212

Q.5. Create two classes DM and DB which store the value of distances. DM stores distances in metres and centimeters and DB in feet and inches. Write a program that can read values for the class objects and add one object of DM with another object of DB.

Use a friend function to carry out the addition operation. The object that stores the results maybe a DM object or DB object, depending on the units in which the results are required. The display should be in the format of feet and inches or metres and cenitmeters depending on the object on display.

Q.6. Create a class rational which represents a numerical value by two double values NUMERATOR & DENOMINATOR. Include the following public member Functions:

- constructor with no arguments (default).
- constructor with two arguments.
- void reduce( ) that reduces the rational number by eliminating the highest common factor between the numerator and denominator.
- Overload + operator to add two rational number.
- Overload >> operator to enable input through cin.
- Overload << operator to enable output through cout.
- Write a main ( ) to test all the functions in the class.

Q.7. Consider the following class definition

```
class father {
protected : int age;
public:
father (int x) {age = x;}
virtual void iam ( )
{ cout < < "I AM THE FATHER, my age is : "<< age<< endl:} };
```

Derive the two classes son and daughter from the above class and for each, define iam() to write our similar but appropriate messages. You should also define suitable constructors for these classes.

Now, write a main() that creates objects of the three classes and then calls iam() for them. Declare pointer to father. Successively, assign addresses of objects of the two derived classes to this pointer and in each case, call iam() through the pointer to demonstrate polymorphism in action.

Q.8. Write a program that creates a binary file by reading the data for the students from the terminal.

The data of each student consist of roll no., name (a string of 30 or lesser no. of characters) and marks.

Q.9. A hospital wants to create a database regarding its indoor patients. The information to store include

(*a*) Name of the patient
(*b*) Date of admission
(*c*) Disease
(*d*) Date of discharge

Create a structure to store the date (year, month and date as its members). Create a base class to store the above information. The member function should include functions to enter information and display a list of all the patients in the database. Create a derived class to store the age of the patients.

List the information about all the patient to store the age of the patients.

List the information about all the pediatric patients (less than twelve years in age).

Q.10. Make a class Employee with a name and salary. Make a class Manager inherit from Employee. Add an instance variable, named department, of type string. Supply a method to to String that prints the manager's name, department and salary. Make a class Executive inherit from Manager. Supply a method toString that prints the string "Executive" followed by the information stored in the Manager superclass object. Supply a test program that tests these classes and methods.

Q.11. Imagine a toll booth with a class called toll Booth. The two data items are a type unsigned int to hold the total number of cars, and a type double to hold the total amount of money collected. A constructor initialises both these to 0. A member function called paying Car ( ) increments the car total and adds 0.50 to the cash total. Another function, called no pay Car

( ), increments the car total but adds nothing to the cash total. Finally, a member function called displays the two totals.

Include a program to test this class. This program should allow the user to push one key to count a paying car, and another to count a non-paying car. Pushing the ESC key should cause the program to print out the total cars and total cash and then exit.

Q.12. Write a function called reversit ( ) that reverses a string (an array of char). Use a for loop that swaps the first and last characters, then the second and next to last characters and so on. The string should be passed to reversit ( ) as an argument.

Write a program to exercise reversit ( ). The program should get a string from the user, call reversit ( ), and print out the result. Use an input method that allows embedded blanks. Test the program with Napoleon's famous phrase, "Able was I ere I saw Elba".

Q.13. Create some objects of the string class, and put them in a Dequeeq - some at the head of the Deque and some at the tail. Display the contents of the Deque using the forEach ( ) function and a user written display function. Then search the Deque for a particular string, using the first That() function and display any strings that match. Finally remove all the items from the Deque using the get Left() function and display each item. Notice the order in which the items are displayed: Using get Left(), those inserted on the left (head) of the Deque are removed in "last in first out" order while those put on the right side are removed in "first in first out" order. The opposite would be true if get Right() were used.

Q.14. Create a base class called shape. Use this class to store two double type values that could be used to compute the area of figures. Derive two specific classes called triangle and rectangle from the base shape. Add to the base class, a member function get_data ( ) to initialise base class data members and another member function display_area ( ) to compute and display the area of figures. Make display_area ( ) as a virtual function and redefine this function in the derived classes to suit their requirements.

Using these three classes, design a program that will accept dimensions of a triangle or a rectangle interactively and display the area.

Remember the two values given as input will be treated as lengths of two sides in the case of rectangles and as base and height in the case of triangles and used as follows:

Area of rectangle = x * y

Area of triangle = ½ * x * y

1

INTRODUCTION TO C++

CHAPTER OUTLINE

1.1 INTRODUCTION

C++ is an "object oriented" programming language created by Bjarne Stroustrup and released in 1985. It implements "data abstraction" using a concept called "classes", alongwith other features to allow object oriented programming. Parts of the C++ program are easily reusable and extensible; existing code is easily modifiable without actually having to change the code. C++ adds a concept called "operator overloading" not seen in the earlier OOP languages and it makes the creation of libraries much cleaner.

> **Concept**
>
> *C++ is a super set of C. we can say, C++ is **increased C** or **C with Classes**.*

C++ maintains aspects of the C programming language, yet has features which simplify memory management. Additionally, some of the features of C++ allow low-level access to memory but also contain high-level features.

C++ could be considered a superset of C. C programs will run in C++ compilers. C uses structured programming concepts and techniques while C++ uses object oriented programming and classes which focus on data.

1.2 FEATURES OF C++

C++ has certain characteristics over other programming languages. The following characteristics are most remarkable :

Reusability

The code written in C++ for one program can be use in another program of C++, Means you can reuse the code written in one in many programs. The concept of inheritance is also provides facility of reusability.

Bottom-Up Approach

C++ uses a bottom-up approach for programming, means it can define functions and data first (or different-different modules) and then integrate them together.

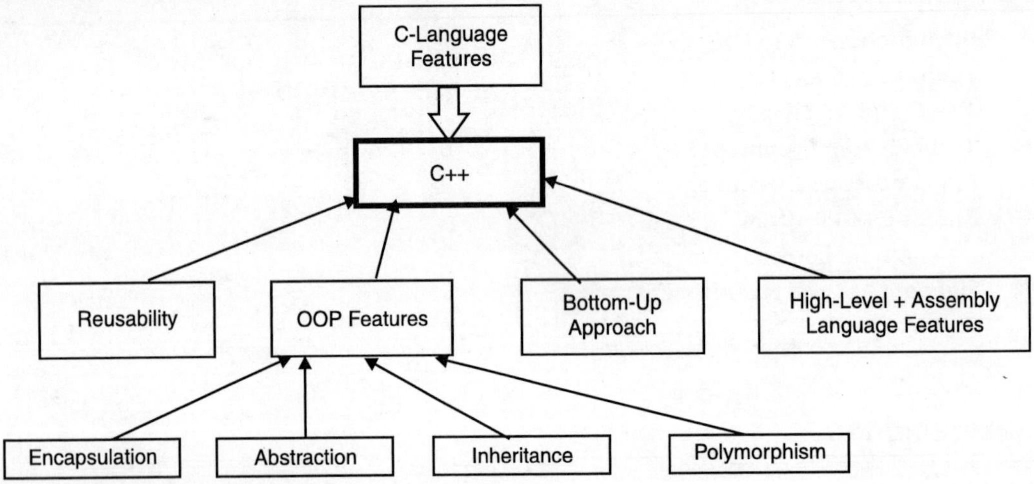

Object Oriented Programming

The object oriented programming provides a mechanism to combine both data and the function of that data into a single unit. Such type of unit is called object. An object's funtions, called member functions in C++, typically provide a way to access the data.

Portability

You can practically compile the same C++ code in almost any type of computer and operating system without hardly making any changes. C++ is the most used and ported programming languages in the world.

Brevity

Code written in C++ is very short in comparison with other languages, since the use of special characters is preferred to key words, saving some effort to the programmer (and extending the life of our keyboards!).

Modular Programming

An application's body in C++ can be made up of several source code files that are compiled separately and then linked together. Saving time since it is not necessary to recompile the complete application when making a single change but only the file that contains it. In addition, this characteristic allows to link C++ code with code produced in other languages, such as Assembler or C.

C Compatibility

C++ is backwards compatible with the C language. Any code written in C can easily be included in a C++ program without hardly making any change.

Speed

The resulting code from a C++ compilation is very efficient, due indeed to its duality as high-level and low-level language and to the reduced size of the language itself.

Related Questions In Exams

Q.1. What do you understand by reusability? How does it help in reducing the cost of software and improving its quality? Discuss the features available in C++ to support reusability.

[MDU-May-2007]

Ans. Reusability : The code written in C++ for one program can be use in another program of C++, Means you can reuse the code written in one in many programs. The concept of inheritance is also provides facility of reusability.

Cost Reducing : The code written in one time can be used in many time and many places. It means saving of several things which affect the cost and improving the quality, such as :

- It will save time for typing the duplicate code in many places,
- It will save human efforts,
- It will reduce the occurrence of errors,
- It will save the memory space consumed by the lengthy code of program,
- It will reduce the length of code.
- Easy maintainable, etc.

Features providing reusability in C++ :

- **Inheritance:** Data and functions can be share in derived classes.
- **Polymorphism:**
  - **(*i*) Function Overloading:** Different functions may use same name for different task.
  - **(*ii*) Function Overriding:** Different classes may have same name functions.

Q.2. What are features of C++ which makes its different from C? *MDU-May-2008]*

Ans.
1. C doesn't support OOP features.
2. C has no classes and Objects.
3. C doesn't support namespaces.
4. C doesn't support operator and function overloading.
5. C doesn't support inline and friend functions.
6. C++ has better dynamic memory management.
7. C++ can use the code of C.
8. C++ may have functions within structure.
9. C++ has better and easy maintenance.
10. C++ has the feature of reusability.

Q.3. What are the basic features C++? *MDU-May-2009]*

Ans. Refer Section 1.2.

1.3 C++ STANDARD LIBRARY

The standard C++ library is a collection of functions, constants, classes, objects and templates that extends the C++ language providing basic functionality to perform several tasks, like classes to interact with the operating system, data containers, manipulators to operate with them and algorithms commonly needed.

C++ Library contains header files and all pre-defined functions such as clrscr(), getch(), strcpy() and etc.

The C++ Standard Library is based upon conventions introduced by the Standard Template Library (STL). Although the C++ Standard Library and the STL share many features, neither is a strict superset of the other. For example, the C++ Standard Library does not provide an slist container.

Header files in the C++ Standard Library do not end in ".h". However, the C++ Standard Library includes 18 header files from the C Standard Library, with ".h" endings. Their use is deprecated. It can be divided into :

- **C++ Standard Library**
  `ios, iostream, iomanip, fstream, sstream`
- **Standard Template Library**
  `vector, deque, list, map, set, stack, queue, bitset, algorithm, functional, iterator`
- **C Standard Library**
  `cassert, cctype, cerrno, climits, clocale, cmath, csetjmp, csignal, cstdarg, cstddef, cstdio, cstdint, cstdlib, cstring, ctime.`

Related Questions In Exams

Q.1. What do you understand by a header file? Discuss the use of iostream header file. *[MDU-Dec-2006]*

Ans. The header files are available in C++ to support all the programs develope by the programmers. The header files containes the definition of pre-defined functions or in-built function. Every program use these header files for accessing the in-built functions. There are several header files supported by C++. Several versions and different C++ products contain a number of different header file according to the provided features. There are some common header files:

- iostream . h (for input/output operations)
- conio . h (for console operations)
- string . h (for string operations)
- math . h (for mathematical operations)
- ctype . h (for character type operations)
- graphics . h (for graphic operations)
- And many more.

All the header files has ". h" extension, and every program contain any header files according to requirement of the program. Every header file used in program with keyword "include" preceded by "#" operator in the top of program. For example :

```
# include < iostream . h>
# include < string . h>
```

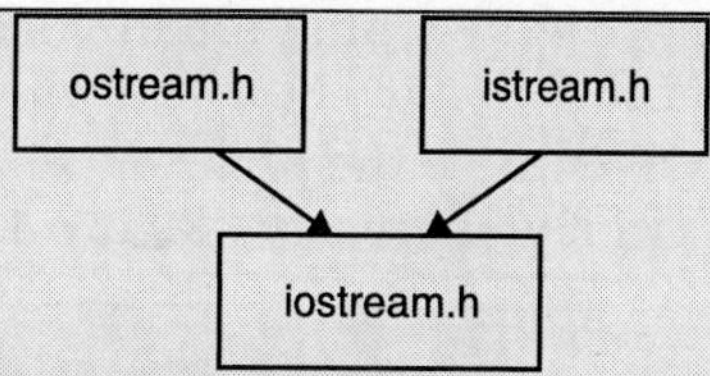

iostream.h :

The iostream.h is a header file which contains input and output function for C++ programs. The iostream.h inherits features of two header files for input and output functions.

The ostream.h contains all functions related to the output on the console, where istream.h contains all functions related to the input form console.

Q.2. What are standard libraries available with C++?

Ans. Refer section 1.3.

1.4 TURBO C++ ENVIRONMENT

Turbo C++ environment is basic IDE for C and C++ Programmer. This is free available in maximum sites on the Web. The basic operation of this environment are given in this section. We have shown how to design a program in Turbo C++ IDE and how to execute them.

1.4.1 Writing a Program

To write a program in Turbo C++ IDE double click on the icon of Turbo C++, available on desktop (if you have installed Turbo C++ with shortcut on desktop). You will see the screen as shown in figure 1.1.

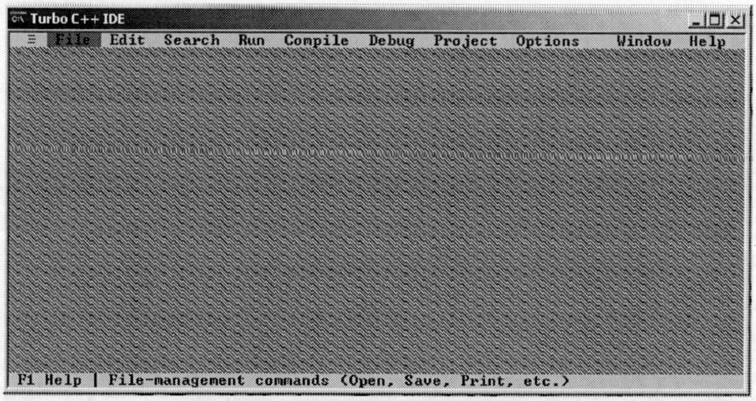

Fig. 1.1. First Screen of Turbo C++

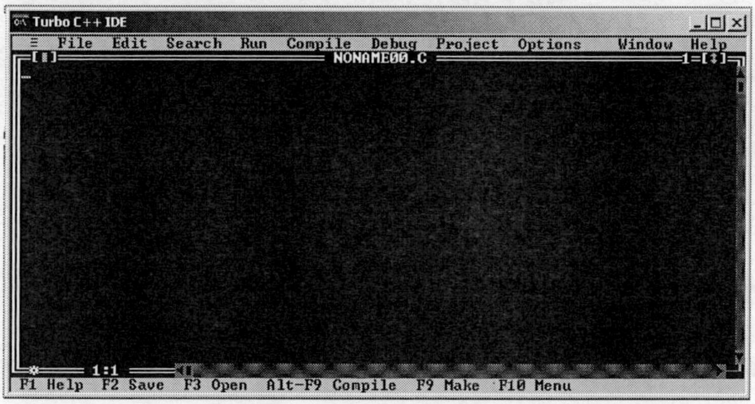

Fig. 1.2. New Window

To write a program in Turbo C++ click on **New** option of **File** menu, on the menubar of Turbo C++ IDE. Now you will see the screen as shown in figure 1.2.

Now type the program whatever you need to write. We type some code of program use them if you are new. Figure 1.3 shows the written code for a first program.

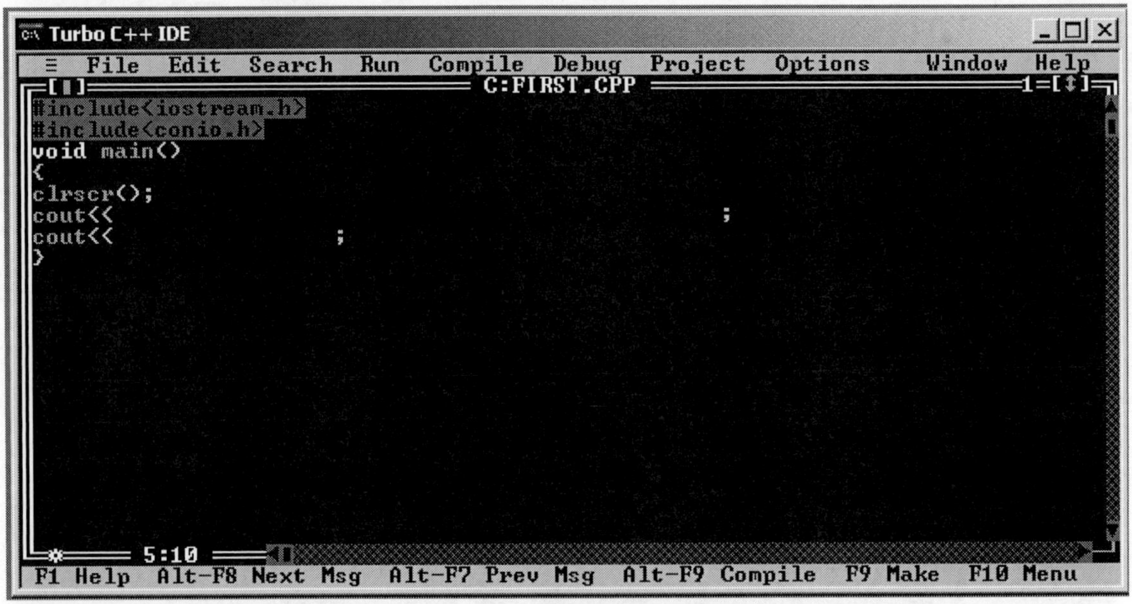

Fig. 1.3. Program in IDE

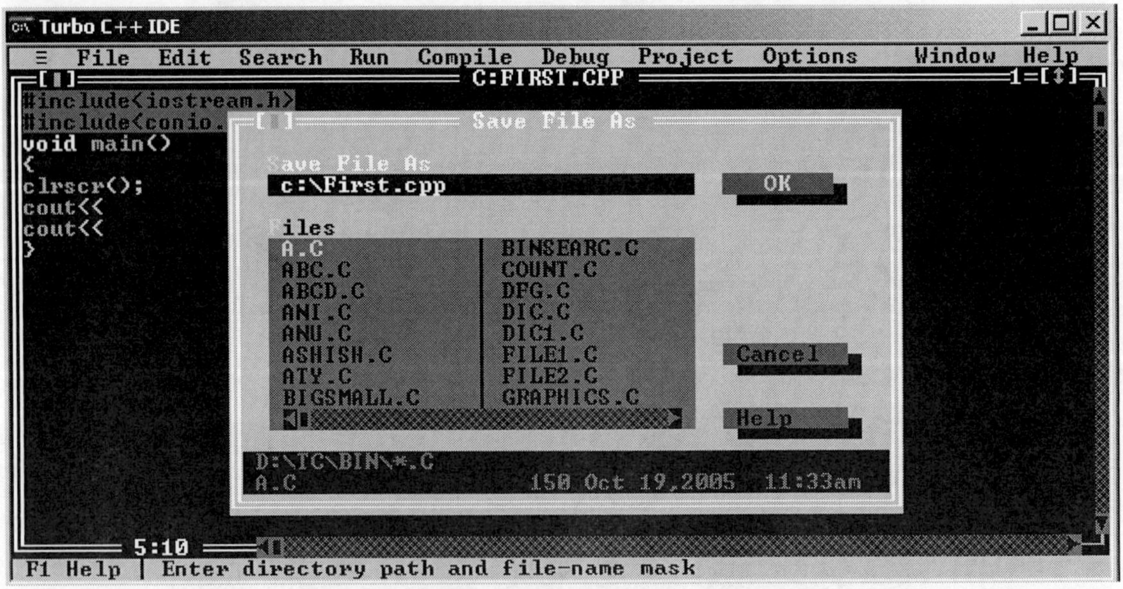

Fig. 1.4. Save Window

Now you can check your program is works or not. But before checking you have to save your program for compilation and execution. So, select the **Save** option from **File** menu of Turbo C++ IDE. IDE will open a save window for saving as shown in figure 1.4.

Then type the name which you need to save, here we have given **"First.cpp".** You must know that "cpp" is an extension for C++ program, all program must save with this extension.

There are several IDEs are available for C++ Programming, such as :

- *Turbo C++,* • *ANSI C++,*
- *Borland C++,* • *Dev C++*

1.3.2 Compilation & Execution

After writing and saving a program you must compile your program. Basically compilation process used to convert high-level language to low-level language, which is executable code for computer. So, to compile the program you have to select the option **Compile** from **Compile** Menu just as figure 1.5.

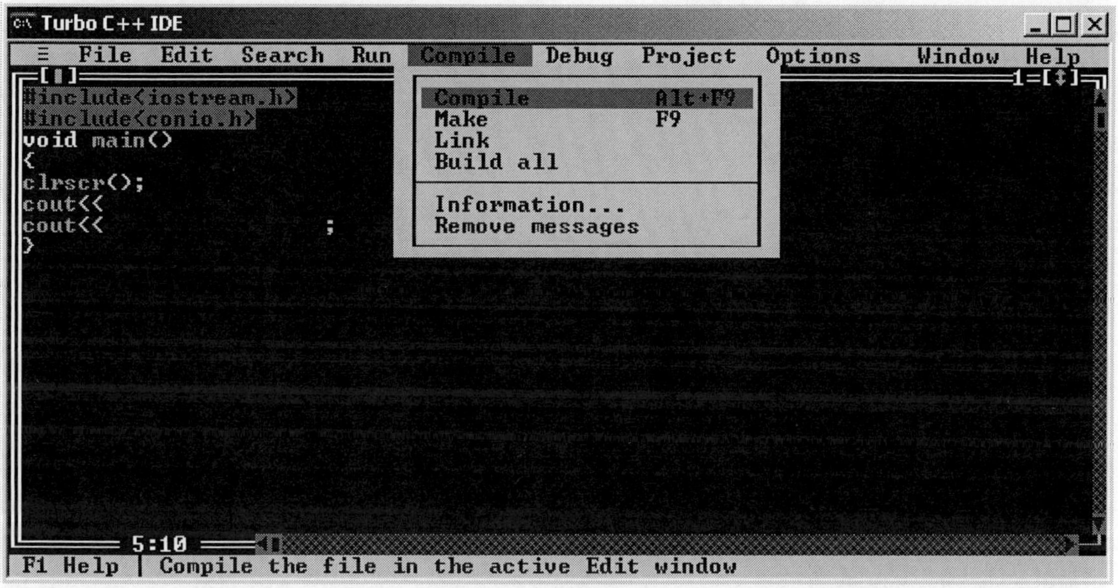

Fig. 1.5. Compilation

When you compile your program, the Turbo C++ compiler convert whole program to executable code, but if it find an error it indicate in the compiling window. Lets see figure 1.6, which indicate zero errors and zero warnings.

If you find any error in the program then you have to edit the program and save it for again compilation, till you get zero errors.

After successful compilation your program is ready for execution, now select **Run** option from Turbo C++ IDE as shown in figure 1.7.

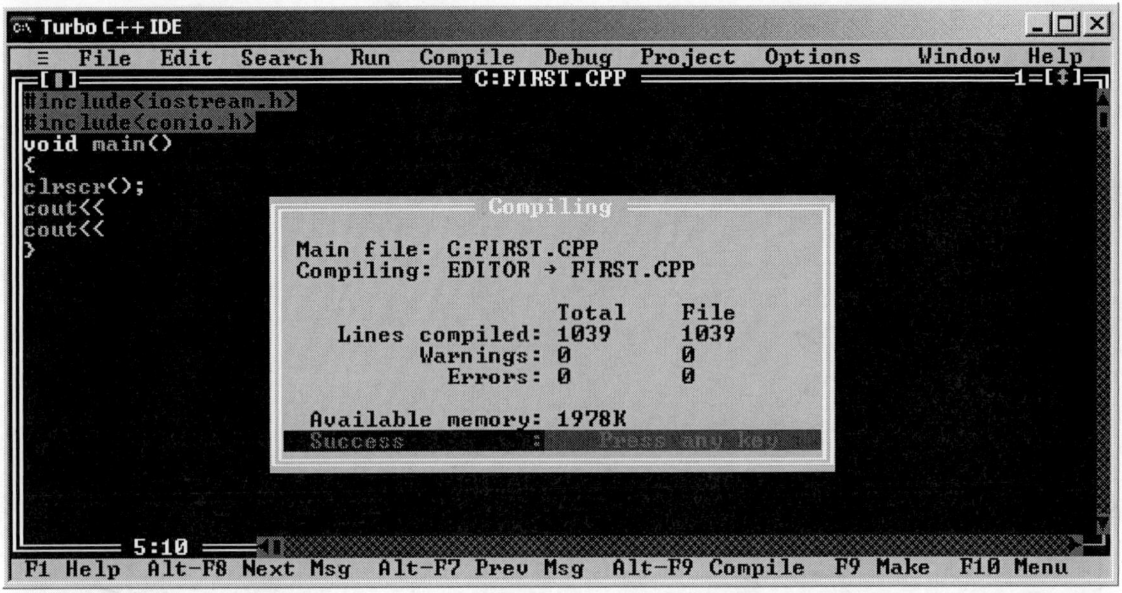

Fig. 1.6. Compilation Window

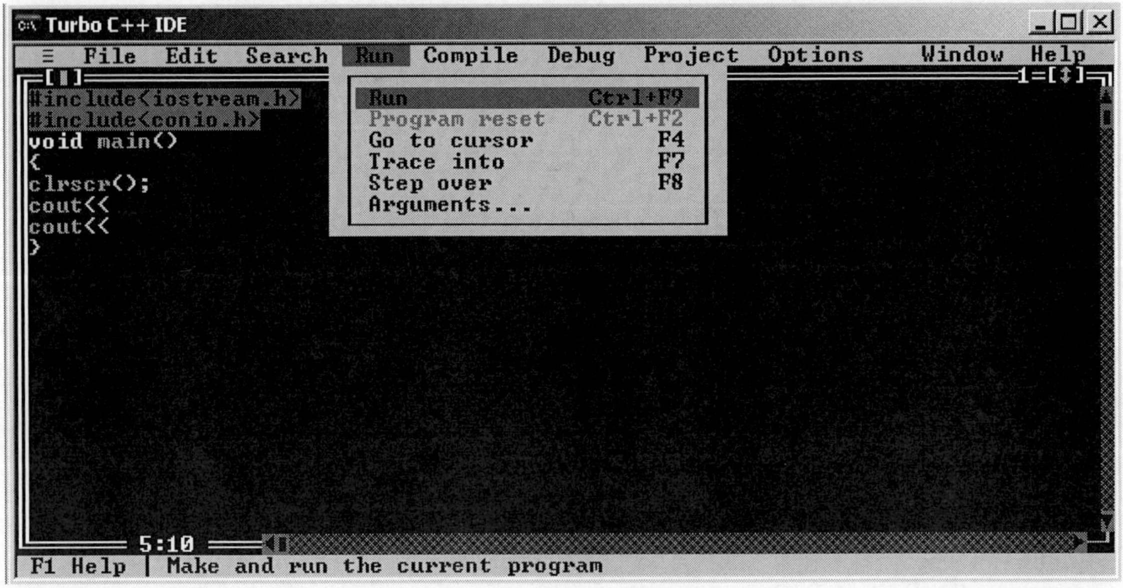

Fig. 1.7. Execution of Program

This Run option will execute your program and display the output on the output window, but to see the output you have to open an **Output** window from the **Window** Menu of Turbo C++ IDE as shown in figure 1.8.

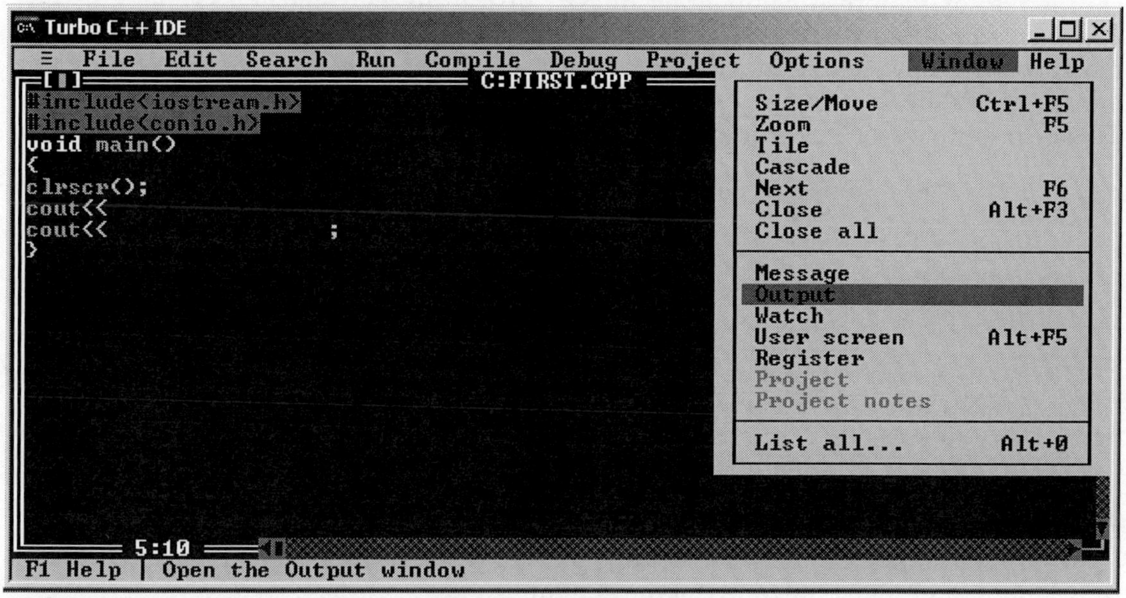

Fig. 1.8. Output Window

This output window show you the output of your program, see figure 1.9.

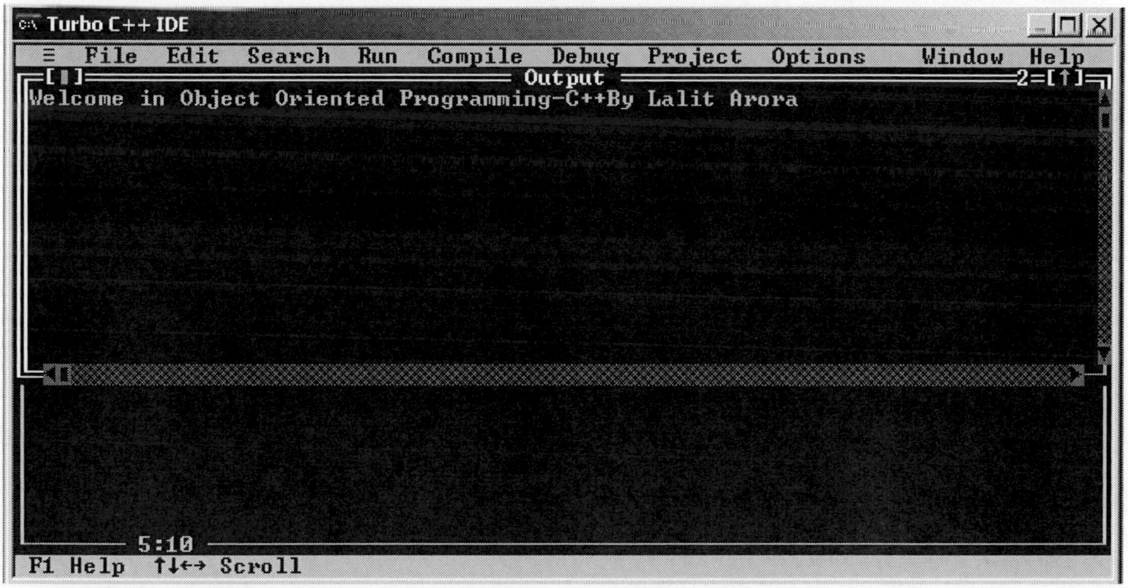

Fig. 1.9. Output Window with output of First.cpp

1.5 PRE-PROCESSORS DIRECTIVES

Pre-processor directives, such as **#include** and **#define**, are typically used to make source programs easy to change and easy to compile in different execution environments. Directives in the source file tell the pre-processor to perform specific actions. For example, the pre-processor can replace tokens in the text, insert the contents of other files into the source file, or suppress compilation of part of the file by removing sections of text. Pre-processor lines are recognised and carried out before macro expansion. Therefore, if a macro expands into something that looks like a pre-processor command, that command is not recognised by the pre-processor.

The pre-processor recognises the following directives :

#define	#error	#import	#undef
#elif	#if	#include	#using
#else	#ifdef	#line	
#endif	#ifndef	#pragma	

The number sign (#) must be the first non-white-space character on the line containing the directive; white-space characters can appear between the number sign and the first letter of the directive.

These are needed for following pupose :

#pragma

This is used to specify diverse options to the compiler. These options are specific for the platform and the architecture of the computer.

#define

This directive replaces occurrence of the defined identifier in the rest of the code used in the whole program by replacement. The replacement an be any expression, a statement, a block of simply anything.

#fdef, #fndef, #if, #endif, #else and #elif

These are the conditional pre-processor directive used to allow to include or discard the part of code in the program.

#line

This pre-processor directive used for debugging the error, when we compile our project which is integrated with many C++ file, it show the line number and filename if error is occurred.

Related Questions In Exams

Q.1. What are pre-processor directives. Why these are needed?
Ans. Refer section 1.4.

1.6 SIMPLE C++ PROGRAM

We are given an example of the simple and classical program "Hello World", named HELLO.CPP, had many interesting parts. This section will review this program in more detail. Example 1.1 produces the original version of HELLO.CPP for your convenience.

Example 1.1 **HELLO.CPP demonstrates the parts of a C++ program.**

```
(line numbers are given for the convenience)
#include <iostream.h>
int main()
```

```
3. {
4. cout << "Hello World!\n";
5. return 0;
6. }
```

OUTPUT

```
Hello World!
```

Line 1, the file iostream.h is included in the file. The first character is the # symbol, which is a signal to the pre-processor. Each time you start your compiler, the pre-processor is run. The pre-processor reads through your source code, looking for lines that begin with the pound symbol (#), and acts on those lines before the compiler runs.

Include is a pre-processor instruction that says, "What follows is a filename. Find that file and read it in right here." The angle brackets around the filename tell the pre-processor to look in all the usual places for this file. If your compiler is set up correctly, the angle brackets will cause the preprocessor to look for the file iostream.h.h in the directory that holds all the H files for your compiler. The file iostream.h (Input-Output-Stream) is used by cout, which assists with writing to the screen. The effect of line 1 is to include the file iostream.h into this program as if you had typed it in yourself.

Line 2, begins the actual program with a function named main(). Every C++ program has a main() function. In general, a function is a block of code that performs one or more actions. Usually functions are invoked or called by other functions, but main() is special. When your program starts, main() is called automatically.

Main(), like all functions, must state what kind of value it will return. The return value type for main() in HELLO.CPP is void, which means that this function will not return any value at all. Returning values from functions is discussed in detail on later chapters.

Lines 3 and 6: All functions begin with an opening brace ({) and end with a closing brace (}). The braces for the main() function are on lines 3 and 6. Everything between the opening and closing braces is considered a part of the function.

Line 4 : The object cout is used to print a message to the screen. The object cout and its related object cin are used in C++ to print strings and values to the screen. A string is just a set of characters.

Here's how cout is used: type the word cout, followed by the output redirection operator (<<). Whatever follows the output redirection operator is written to the screen. If you want a string of characters written, be sure to enclose them in double quotes ("), as shown on line 4.

Line 5: All ANSI-compliant programs declare main() to return an int. This value is "returned" to the operating system when your program completes. Some programmers signal an error by returning the value 1. In this guide, main() will always return 0.

The main() function ends on line 6 with the closing brace.

A Brief Look at cout

Here we are given a brief look at cout without fully understanding how it works. To print a value to the screen, write the word cout, followed by the insertion operator (<<), which you create by typing the less-than character (<) twice. Even though this is two characters, C++ treats it as one.

Follow the insertion character with your data. Example 1.2 illustrates how this is used. Type in the example exactly as written, except substitute your own name where you see Ankit Gupta (unless your name is Ankit Gupta.).

Example 1.2 **Using cout**

```
(line numbers are given for the convenience)
1. // Example 2.2 using cout
2. #include <iostream.h>
3. int main()
4. {
5. cout << "Hello there.\n";
6. cout << "Here is 5: "  << 5 << "\n";
7. cout << "The manipulator endl writes a new line to the screen."
   << endl;
8. cout << "Here is a very big number:\t" << 70000 << endl;
9. cout << "Here is the sum of 8 and 5:\t" << 8+5 << endl;
10. cout << "Here's a fraction:\t\t" << (float) 5/8 << endl;
11. cout << "And a very very big number:\t" << (double) 7000 * 7000
    << endl;
12. cout << "Don't forget to replace Jesse Liberty with your
    name...\n";
13. cout << "Ankit Gupta is a C++ programmer!\n";
14. return 0;
15. }
```

OUTPUT

```
Hello there.
Here is 5: 5
The manipulator endl writes a new line to the screen.
Here is a very big number:          70000
Here is the sum of 8 and 5:         13
Here's a fraction:                  0.625
And a very very big number:         4.9e + 07
Don't forget to replace Jesse Liberty with your name...
Ankit Gupta is a C++ programmer!
```

The statement #include <iostream.h> causes the iostream.h file to be added to your source code. This is required if you use cout and its related functions.

On line 5 is the simplest use of cout, printing a string or series of characters. The symbol \n is a special formatting character. It tells cout to print a newline character to the screen.

Three values are passed to cout on line 6, and each value is separated by the insertion operator. The first value is the string "Here is 5: ". Note the space after the colon. The space is part of the string. Next, the value 5 is passed to the insertion operator and the newline character (always in double quotes or single quotes). This causes the line Here is 5: 5 to be printed to the screen. Because there is no newline character after the first string, the next value is printed immediately afterwards. This is called concatenating the two values.

On line 7, an informative message is printed, and then the manipulator endl is used. The purpose of endl is to write a new line to the screen.

On line 8, a formatting character, \t. This inserts a tab character and is used on lines 8 – 12 to line up the output. Line 8 shows that not only integers, but long integers as well can be printed. Line 9 demonstrates that cout will do simple addition. The value of 8 + 5 is passed to cout, but 13 is printed.

> *When you Compile .CPP file compiler generates new .OBJ file, and after execution .BAK and .EXE files are generated.*

On line 10, the value 5/8 is inserted into cout. The term (float) tells cout that you want this value evaluated as a decimal equivalent, and so a fraction is printed. On line 11 the value 7000 * 7000 is given to cout, and the term (double) is used to tell cout that you want this to be printed using scientific notation.

On line 14, you substituted your name, and the output confirmed that you are indeed a C++ programmer. It must be true, because the computer said so!

1.7 HEADER FILES AND NAMESPACES

The final version of the C++ standard defines names from system header files in a "namespace" called std. The standard requires that you specify which names in a standard header file you'll be using. While most compilers now accept the new format, they don't require it. Subsequent versions will make this syntax mandatory, so you should begin using it now. The Deitel textbook conforms to the standard for C++ header files, but not for C header files.

As an example, the following code was formerly acceptable C++ but is invalid under the latest standard :

```
#include <iostream.h>
int main()
{
    cout << "Hello, world!" << endl;
}
```

The following three examples show how the **C++ standard** would now represent this code.

// Option 1

```
#include <iostream>
int main()
{
    std::cout << "Hello, world!" << std::endl;
}
```

// Option 2

```
#include <iostream>
using std::cout;
using std::endl;
int main()
```

```
{
    cout << "Hello, world!" << endl;
}
```

// Option 3

```
#include <iostream>
using namespace std;
int main()
{
    cout << "Hello, world!" << endl;
}
```

For all of these, notice that the ".h" extension on the name of the system header file is omitted. You would still include the ".h" extension for any header file you write, but the extension is omitted for system header files. The second option above is the preferred option, and is the one that is used in the Deitel textbook. Don't include **using** directives in your header files unless the name you are specifying is actually used in the header file. **In most cases the using directive will appear in the implementation file.**

When you include a system header file, you should indicate which specific names are defined in the header file that you will be using in your program. If you are including multiple system header files, you should put the appropriate **using** directives after each header files.

```
#include <iostream>
using std::cout;
using std::endl;

#include <iomanip>
using std::setw;

#include <cstring>
using std::strcpy; // Omit this line for Visual C++
using std::strcat; // Omit this line for Visual C++

#include <cassert>
using std::assert; // Omit this line for Visual C++

#include "string1.h"
```

Put C++ system header files with their **using** directives first. Any C header files come next. Notice that header files that are hold-overs from the C language begin with the letter "c" and omit the ".h" extension. The header file formerly referred to as <assert.h> is now <cassert>. Microsoft Visual C++ (version 6 and earlier) doesn't allow **using** directives with C header files, but this problem is fixed started with version 7.

Related Questions In Exams

Q.1. What do you understand by a namespace? Discuss the syntax for defining a namespace. Discuss its using suitable example. *[MDU-May-2007]*

Ans. Namespace : namespace is a new concept introduced from the ANSI C++, which can be used as alternate of header files. For example std namespace can be used instead of iostream.h. Basically name space is a collection of identifiers such as variables, structures, classes and their functions. In ANSI C++, there are several pre-defined namespaces.

The namespace can be used with using keywords, like as follows :

```
using namespace std;
```

User also can define their new namespaces, using following syntax :

```
Syntax :
    namespace <name-space name>
    {
        //declaration of variables, functions, classes, etc.
    }
Example :
    namespace data
    {
        class mydata
        {
            char My_name[10];
            int My_age;
        public :
            void get_data()
            {
                // statements
            }
            void putdata()
            {
                //statements
            }
        }
    }
```

Q.2. What are the header files & library files? Why these are needed?

Ans. The header files are available in C++ to support all the programs develope by the programmers. The header files containes the definition of pre-defined functions or in-built function. Every program use these header files for accessing the in-built functions. There are several header files supported by C++. Several versions and different C++ products contain a number of different header file according to the provided features. There are some common header files:

- iostream . h (for input/output operations)
- conio . h (for console operations)
- string . h (for string operations)
- math . h (for mathematical operations)
- ctype . h (for character type operations)
- graphics . h (for graphic operations)
- And many more.

All the header files has ". h" extension, and every program contain any header files according to requirement of the program. Every header file used in program with keyword "include" preceded by "#" operator in the top of program. For example :

```
# include < iostream . h>
# include < string . h>
```

Header Files : The header files are necessary file require for the all programs of C++ for input/output and other important functions of the program. **Examples :** iostream.h, ostream.h, istream.h.

Library Files : The Library files are those file which are used for some specific task. These files contain some related functions. **Example :** string.h, math.h, graphics.h.

Summary

- C++ is a superset of C. We can say, C++ is increased C or C with classes.
- C++ standard library contain huge number of functions which help to develop the programs.
- Pre-processors directives such as #include and #define, are typically used to make source programs easy to change and easy to compile in different execution environment.

Solved Problems

Q. 1.1. What are the types of computer languages?

Ans. Computer languages basically two types :

(1) High-level languages

(2) Low-level languages

1. High Level Languages: The high-level languages are very closed to humans, but not understandable by the computers. These type of languages are made in the natural languages (such as English, French, Hindi). The example of these type of languages are C, C++, PASCAL, COBOL, FORTRAN etc.

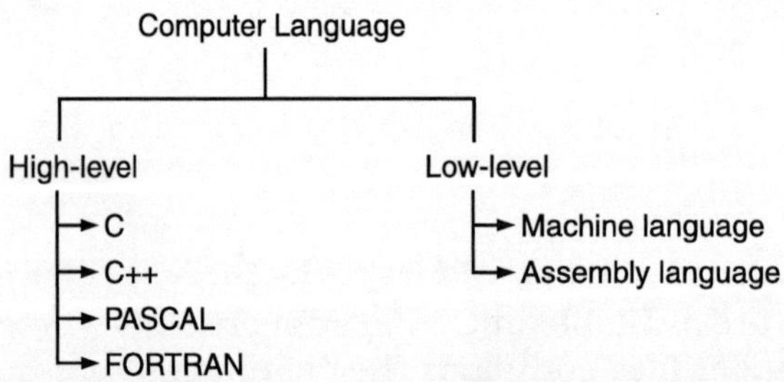

2. Low Level Languages: This type of languages are very close to computers and easily under standable by the computers. There are several name are used for these type of languages such as machine language, executable code and machine code, etc. All the high level languages programs are firstly translated into the low-level languages or machine code than the program will be execute. There are several translator are available for translating the high-level program to the executable code.

Middle Level Languages: If any high-level language uses the properties of low-level language, then it is called middle level language like C language, uses the features of assembly language, therefore it also called middle level language.

Q. 1.2. What do you mean by Translator?

Ans. In the computer we are using high-level language to develope our software and programs, which uses the words of natural language. But computer can't understand this type of language. Thus translators are used to convert programs from high-level to low-level language.

There are three types of Translators :

(1) Compiler

(2) Interpreter

(3) Assembler

1. Compiler: The compiler translate the high-level program to low-level code. It translate whole program at once.

2. Interpreter: The interpreter translate the code from high-level language to machine level. But it translate whole program statement by statement. One statement at a time will convert to machine code.

3. Assembler: This is also a translator, convert an assembly program to machine code.

Q. 1.3. What is an algorithm and a program ? And what is the difference between them ?

Ans. Algorithm: Algorithm is a step-by-step solution for a problem. It provides the method to solve the problem. From the algorithm you can understand how can be solve a particular problem.

Program: Program is set of instruction written in any computer language. Program is an implementation of algorithm on computer to solve the problem. Program is only way one computer to find the result or solution of a particular problem.

Q. 1.4. Why we use comment? While it is not executable statements.

Ans. C++ provides the facility of comment in programs. There are two way of comment in C++.

(1) Using /* .... */ called multiple line comment.

(2) Using // called single line comment.

For example : **Multiple Line Comment :**

```
/* It is my First Program in C language.
                Name : Sunil Kumar
                Date : 10-10-05 */
```

Single Line Comment:

```
// It is my first program in C-language.
// Name : Sunil Kumar
// Date : 10-10-05.
```

The basic use of comment is easy understanding of program and statement of function.

At the time of software maintenance we required to know that what task performed by any particular function or statement and why ?

Therefore at the time of software development we must comment for each function and any logical statement or wherever is required.

Q. 1.5. What are the rules for identifiers ? How many character can be include in an identifier name?

Ans. Rules for Identifiers:

1. First character of identifier name should be alphabet.
2. Any keyword cannot be used as a name of identifier.
3. Only alphabetic character, digits and under scores are permitted for the name of identifier.
4. Upper case and lower case letters are distinct, *i.e.*, *X* and *x* are different valid identifiers in C++.

→ ANSI C++ has no limit on character length for identifier names.

Q. 1.6. What do you meaning of Typecasting ?

Ans. Typecasting mean conversion of data types from one to another. It is most important for mathematical expression in the computers. If an integer variable used with any float variable in any mathematical expression, then this integer variable required to convert into the float type. There are two version of type casting.

1. Total = (float) *a* + *b*
2. Total = float (*a*) + *b*

Q. 1.7. What is an expression ?

Ans. An expression is a statement in which mathematical operators are used. Actual this expression shows some operations on the operands. Both side of an operators operands are available. There are some examples of expressions :

```
Total = marks 1 + marks 2 ;
C = a – b ;
F = a * b * c + (a – d)/2 ;
```

Q. 1.8. What are the header files?

Ans. The header files are available in C++ to support all the programs develope by the programmers. The header files containes the definition of pre-defined functions or in-built function. Every program use these header files for accessing the in-built functions. There are several header files supported by C++. Several versions and different C++ products contain a number of different header file according to the provided features. There are some common header files:

- iostream . h (for input/output operations)
- conio . h (for console operations)
- string . h (for string operations)
- math . h (for mathematical operations)
- ctype . h (for character type operations)
- graphics . h (for graphic operations)
- And many more.

All the header files has ". h" extension, and every program contain any header files according to requirement of the program. Every header file used in program with key word "include" preceded by "#" operator in the top of program. For example :

```
# include < iostream . h>
# include < string . h>
```

Q. 1.9. What is the method of execution of C++ program?

Ans. We are explaining the Turbo C++ Execution Method:

You must complete before the execution of the program. There are two methods of the program execution :

1. Menu Driven
2. Using shortcut keys.

1. Menu Driven : After the completion of program do the following menu driven steps:

Run ® Execute

2. Shortcut Keys :

(*i*) Alt + F9 : Compile
(*ii*) Ctrl + F9 : Execute
(*iii*) Ctrl + F5 : Output window

Q. 1.10. What is cout and cin?

Ans. In the C++ cout and cin are the objects defined in *iostream* file. Basically cout used for output and cin used for input of data. Cout is the object of *ostream* class and cin is the object of *istream* class in the iostream file. Both are used with output (<<) and input (>>) operators respectively with *cout* and *cin*. We can also call some function with these objects.

For example :

```
# include <isotream.h>
int main ()
    {
    cout. precision (4);
    cout. width (10);
    cout << 10.12345 << endl; // displays 10.12

    cout. fill ('*');
    cout. width (10);
    cout << 10.12345 << endl; // displays ***** 10.12
    // field width applies to strings, too
    cout. width (10);
    cout << "Hi" << endl; // displays ******* Hi!
    cout. width (10);
    cout. set (ios : : left); // left justify
    cout/ << 10.12345; // displays 10.12 *****
    return 0;
}
```

OUTPUT:

```
10.12
***** 10.12
*******Hi
10.12*****
```

Program without Classes

Q. 1.11. Write program to input alphabet character and print its predecessor and sucesor character.

Ans.

```
#include<iostream.h>
  void main ()
    {
        char ch;

        cout << "In Enter any character;" << endl;
        cin >> ch;
        cout << "\n the predecessor character is:" << ch - 1;
        cout << "\n the sucessor character is:" << ch + 1;
        }
```

Q. 1.12. Write a program to input an alphabet character in lower case and print in an upper case or *vice-versa*.

Ans.

```
# include<iostream.h>
void main ()
    {
        char ch;
        cout <<"\n Enter Any character:" << endl;
        cin >> ch;
        if (ch > = 65 && ch < = 90)
                          // ASCII value for A to Z  = 65 to 90
                 ch = ch + 32;
        else if (ch > = 97 && ch < = 122)
                          // ASCII value for a to z = 97 to 122
                 ch = ch - 32;
        cont <<"\n second case of character" << ch;
        }
```

Q. 1.13. Write a program to input two integers and interchange the value without using third variable.

Ans.
```
#include<iostream.h>
#include<conio.h>
void main ()
{
    int a, b;
    cout <<"\n Enter two values:",
    cin >> a >> b;
    cout << "value of a = "<<a <<" and b = "<<b;
    a = a + b;
    b = a - b;
    a = a - b;
    cout << "/n After swaping:";
    cout << "\n value of a =" << a;
    cout << "\n value of b = "<< b;
    }
```

OUTPUT:

```
Enter Two values:
10
15
Value of a = 10 and b = 15
After swaping
Value of a = 10
Value of b = 15
```

Q. 1.14. Write program to input a number and print all digits in words.

Ans.
```
#include <iostream.h>
void main ()
{
    int num, a, b.,

    cout <<"\n Enter the number;" << endl;
    cin >> num;
    b = 0;
    while (num! = 0)
        {
        a = num % 10;
        b = b * 10 + a;
        num = num/10;
        }
        num = b;

    while (num! = 0)
        {
        a = num % 10;
        num = num/10;
        switch (a)
            {
                case 1 : cout <<"One";
                            break;
```

```
                    case 2 : cout <<"Two";
                             break;
                    case 3 : cout <<"Three" ;
                             break;
                    case 4 : cout <<"Four";
                             break;
                    case 5 : cout <<"Five";
                         break;
                    case 6 : cout <<"Six";
                             break;
                    case 7 : cout <<"Seven";
                         break;
                    case 8 : cout <<"Eight";
                         break;
                    case 9 : cout <<"Nine";
                         break;
                    case 10 : cout <<"Zero";
                          break;
                    }
                }
            }
```

OUTPUT:

```
Enter the number:
346
Three four six.
```

Q. 1.15. Write a program to display the series of numbers given below where *n* is a row number input by user. (input *n* = 4)

(a)	**(b)**	**(c)**
1	**1**	**1**
1 2	**2 1**	**2 5**
1 2 3	**3 2 1**	**3 6 8**
1 2 3 4	**4 3 2 1**	**4 7 9 10**

Ans.

```
#include < iostream.h>
  void main ()
  {
      int n;
      cout <<"/n Enter the row number;" endl;
      cin >> n;
      for (int i = 1; i < = n; i + r)
          {
          for (int j = 1, j < = i, j ++)
              {
                  cout << j << "\t";
                      }
              cout << endl;
              }
          }
```

OUTPUT:

```
Enter the row number:
4
1
1 2
1 2 3
1 2 3 4
```

(*b*)
```
#include<iostream.h>
  void main()
  {
      int n;
      cout<<"\n Enter the row number << endl;
      cin >> n;
      for (int i = 1; i < = n; i ++)
          {
          for (int j = i; j > = 1; j - -)
              {
                  cout << j <<"\t";
                  }
              cout << endl; } }
```

OUTPUT:

```
Enter the row number
5
1
2 1
3 2 1
4 3 2 1
5 4 3 2 1
```

(*c*)
```
#include <iostream.h>
  void main ()
  {
      int n, m = 1;

      cout << "\n Enter the row number:" << endl;
      cin >> n;

      for (int i = 1; i < = n; i ++)
          {
              x = 3;
      for (int j = 1; j < = i, j ++)
          {
              if (j = = 1)
                  cout << m;
              else
              {cout << m + x;
              x --
                  }
          }
          m ++;
          cout << endl;
          }
      }
```

OUTPUT:

```
Enter the row number:
4
1
2  5
3  6  8
4  7  9  10
```

Q. 1.16. Write a program to display the sum of first ten even natural number square (e.g., $2^2 + 4^2 + 6^2 + 8^2 + ...$)

Ans.
```
#include <iostream.h>
void main ( )
{
    int sum = 0, even = 2;
    for (int i = 1; i < = 10; i ++)
        {
            sum = sum + even * even;
            even + = 2;  //even = even + 2;
            }
        cout <<"\n sum = " << sum;
        }
```

Q. 1.17. Write a program to input number and check it is prime or not.

Ans.
```
#include <iostream.h>
void main ()
{
    int num, p = 1;
    cout <<"\n Enter the number;";
    cin >> num;
    for (int i = 2; i < = num/2; i ++)
        {
            if (num % i = = 0)
                p = p * 0;
        else
            p = p * 1;
        } }
        if (p ! = 0)
        cout << "\n number is prime";
    else
        cout << "\n number is not prime;
    }
```

OUTPUT

```
Enter the number
6
number is not prime.
```

Q. 1.18. Write a program to print series of prime number upto given number.

Ans.
```
#include <iostream.h>
void main ()
{
    int n;
    cout << "\n Enter the ring of prime number;" << endl;
    cin >> n;
    cout << "\n The series of prime Numbers;" << endl;
    for (int i = 1; i < = n; i ++)
        { p = 1;
        for (int j = 2; j < = i/2; j ++)
            {
                if (i% j = = 0)
                        p = p * 0;
                else
                        p = p*1;
```

```
                }
                if (p = = 1)
                        cout << i << "/t";
                }
        }
```

OUTPUT

```
Enter the range of prime number:
10
The series of prime numbers:
2  3  5   7   11   13   17   19   23   29.
```

Q. 1.19. Write a program to display the Fibbnacci series: (0, 1, 1, 2, 3, 5, 8, 13 ....)

Ans.

```
#include <iostream.h>
void main ()
{
    int fib, f1, f2, n;
    f1 = 0
    f2 = 1
    cout << "\n Enter the range;" endl; cin >> n;
    cout <<f1 << "\t" << f2;
    cout << "\n The seriese of fibacci series" << endl;
    for (int i = 2; i < = n; i ++)
        {
            fib = f1 + f2;
            cout << fib << "\t";
            f1 = f2;
            f2 = fib;
        }
    }
```

OUTPUT

```
Enter the range
10
The series of fibonacci series:
0   1   1   2   3   5   8   13   21   34
```

Q. 1.20. Write a program of input a name and find the length of name.

Ans.

```
#include <iostream.h>
#include <string.h>
void main ()
{
    char name [20];
    int len;
    cout << "\n Enter your name" ;
    cin >> name;
    // using string function
    // len =strlen (name);
    // without using string function
    for (int len = 0; name [len]! = '\0' len ++);
    cout << "\n Length of name = " << len;
    }
```

Q. 1.21. What does #include do?

Ans. This is a directive to the pre-processor, which runs when you call your compiler. This specific directive causes the file named after the word include to be read in, as if it were typed in at that location in your source code.

Q. 1.22. What differentiates a good comment from a bad comment?

Ans. A good comment tells the reader why this particular code is doing whatever it is doing or explains what a section of code is about to do. A bad comment restates what a particular line of code is doing. Lines of code should be written so that they speak for themselves. Reading the line of code should tell you what it is doing without needing a comment.

Q. 1.23. Why use unnecessary parentheses when precedence will determine which operators are acted on first?

Ans. Although it is true that the compiler will know the precedence and that a programmer can look up the precedence order, code that is easy to understand is easier to maintain.

Q. 1.24. If the relational operators always return 1 or 0, why are other values considered true?

Ans. The relational operators return 1 or 0, but every expression returns a value, and those values can also be evaluated in an if statement. Here's an example:

```
if ((x = a + b) == 35)
```

This is a perfectly legal C++ statement. It evaluates to a value even if the sum of a and b is not equal to 35. Also note that x is assigned the value that is the sum of a and b in any case.

Q. 1.25. What effect do tabs, spaces, and new lines have on the program?

Ans. Tabs, spaces, and new lines (known as whitespace) have no effect on the program, although judicious use of whitespace can make the program easier to read.

Q. 1.26. Are negative numbers true or false?

Ans. All non-zero numbers are positive and negative, are true.

Q. 1.27. Why not make all variables global?

Ans. There was a time when this was exactly how programming was done. As programs became more complex, however, it became very difficult to find bugs in programs because data could be corrupted by any of the functions—global data can be changed anywhere in the program. Years of experience have convinced programmers that data should be kept as local as possible, and access to changing that data should be narrowly defined.

Q. 1.28. How do you choose between if/else and switch?

Ans. If there are more than just one or two else clauses, and all are testing the same value, consider using a switch statement.

Q. 1.29. How do you choose between while and do...while?

Ans. If the body of the loop should always execute at least once, consider a do...while loop; otherwise, try to use the while loop.

Q. 1.30. How do you choose between while and for?

Ans. If you are initialising a counting variable, testing that variable, and incrementing it each time through the loop, consider the for loop. If your variable is already initialised and is not incremented on each loop, a while loop may be the better choice.

Q. 1.31. Is it better to use while (1) or for (;;)?

Ans. There is no significant difference.

Q. 1.32. How does a function declaration is differ from a function definition?

Ans. Function declaration includes name of function, return type, number of arguments and type of arguments, while the function definition includes all these items and the operation on these arguments. For example:

```
int multiplication(int, int); // function declaration
// function definition
int multiplication (int a, int b)
{
    return (a * b);
}
```

Q. 1.33. Why aren't changes to the value of function arguments reflected in the calling function?

Ans. Arguments passed to a function are passed by value. That means that the argument in the function is actually a copy of the original value.

Q. 1.34. How do you choose between recursion and iteration?

Ans. Some problems cry out for recursion, but most problems will yield to iteration as well. Put recursion in your back pocket; it may come in handy some Chapter.

Q. 1.35. Explain what a reference variable is?

Ans. A reference variable in C++ is a new type that provides the means by which an alias (an another name of someone) to another variable can be created. This means that whenever the alias is used, the action is really taking place on the object to which the alias refers. While a reference variable is conceptually similar to a pointer variable in that they both refer, or point, to some other object, a reference variable is a "hidden" pointer because unlike a pointer variable, it does not have its own unique address.

For example:

```
int num;
int& number = num;
```

Here number is a reference variable for num. The advantage of reference variable is that in many case they can replace function calls that are "paas by address" with calls that are "pass by reference". This syntax does not involve your having to manually generate the address of a variable, nor do you have to create a pointer variable in which to receive the address.

Q. 1.36. What is the difference between malloc() and *new* operator?

Ans. Like the malloc() function in C, the *new* operator in C++ is used to allocate contiguous, unnamed memory at execution time. Unlike malloc(), however, the *new* operator no longer needs to use the *size of* keyword to specify the exact number of bytes needed. The fact that different types occupy different amounts of storage is handled automatically by the compiler.

Q. 1.37. What happens if I write to element 25 in a 24-member array?

Ans. You will write to other memory, with potentially disastrous effects on your program.

Q. 1.38. What is in an uninitialised array element?

Ans. Whatever happens to be in memory at a given time. The results of using this member without assigning a value are unpredictable.

Q. 1.39. Can I combine arrays?

Ans. Yes. With simple arrays you can use pointers to combine them into a new, larger array. With strings you can use some of the built-in functions, such as strcat, to combine strings.

Q. 1.40. Write a function to return the sum of digits of given number.

Ans.

```
#include <iostream.h>
int sumdigit(int);
void main ()
```

```
{
     int num, sum;
     cout<<"\n Enter the number:";
     cin>>num;
     sum = sumdigit(num);
     cout<<"\n Sum of digits :"<<sum;
}
int sumdigit(int n)
{
     int s;
     while(n != 0)
     {
          s = s + (n % 10);
          n = n / 10;
     }
     return (s);
}
```

OUTPUT:

```
Enter the number:
123
Sum of digits:
6
```

Q. 1.41. Write a program to calculate the area of circle, rectangle, and triangle using functions.

Ans.

```
#include<iosteram.h>
void menu();
void area_circle();
void area_rect();
void area_tri();
void main()
{
     int option;
do {

menu();
cin>>option;
swtich(option)
{
     case 1:
          area_circle();
          break;
     case 2:
          area_rect();
          break;
     case 3:
          area_tri();
          break;
}
```

```
}while(option > 3);
}
void menu()
{
     cout<<"\n ***** Menu ****** ";
     cout<<"\n 1   :    Area of Circle";
     cout<<"\n 2   :    Area of Rectangle";
     cout<<"\n 3   :    Area of Triangle";
     cout<<"\n 4 or greater : Exit";
     cout<<"\n Enter your Choice   :";
}
void area_circle()
{
     int r;
     cout<<"\n Enter the radius :";
     cin>>r;
     float area = 3.14*r*r/4;
     cout<<"\n Area of circle : "<<area;
}
void area_rect()
{
     int a,b;
     cout<<"\n Enter the sides of rectangle :";
     cin>>a>>b;
     int area = a*b;
     cout<<"\n Area of Rectangle : "<<area;
}
void area_triangle()
{
     float s,a,b,c;
     cout<<"\n Enter the sides of triangle :";
     cin>>a>>b>>c;
     s = (a+b+c) /2;
     float area = s*(s-a)*(s-b)*(s-c);
     cout<<"\n Area of triangle : "<<area;
}
```

OUTPUT:

```
***** Menu ******
1   :    Area of Circle
2   :    Area of Rectangle
3   :    Area of Triangle
4 or greater : Exit
Enter your Choice  :
1
Enter the radius :
4
Area of Circle     :
```

```
12.56
***** Menu ******
1 : Area of Circle
2 : Area of Rectangle
3 : Area of Triangle
4 or greater : Exit
Enter your Choice  :
5
```

Q. 1.42. Write a program to read an integer and evaluate it's factorial by recursion.

Ans.
```
#include<iostream.h>
int factorial(int);
void main()
{
    int num, fact;
    cout<<"\n Enter the number :";
    cin>>num;
    fact = factorial(num);
    cout<<"\nFactorial of number is :"<<fact;
}
int factorial(int n)
{
    if (n==0)
        return(1);
    else
        return( n*factorial(n-1));
}
```

OUTPUT:

```
Enter the number:
5
Factorial of number is:
120
```

Q. 1.43. Write a program to display fibonacci series using recursion.

Ans.
```
#include<iostream.h>
int fibo(int);
void main()
{
    int i;
    for(i=1; i<=10; i++)
    cout<<fibo(i);

}
int fibo(int n)
{
    if (i==1)
        return(0);
    else if(i==2)
        return(1);
```

```
else
    return(fibo(i-1) + fibo(i-1));
}
```

Q. 1.44. Write a program to display the nth number of the fibonacci series using recursion.

Ans.
```
#include<iostream.h>
int fibonum(int);
void main()
{
    int p,num;
    cout<<"\nEnter the position of element you want from fibonacci
    series:";
    cin>>p;
    num = fibonum(p);
    cout<<"\n Number is : "<<num;
}
int fibonum(int x)
{
    if(x==1 || x==0)
        return (x);
else
        return(fibonum(x-1) + fibonum(x-2));
}
```

OUTPUT:

```
Enter the position of element you want from fibonacci series :
4
Number is :
3
```

Q. 1.45. Write a program to display reverse of an integer number using recursion.

Ans.
```
#include<iostream.h>
int reverse(int);
void main()
{
    int n, rev;
    cout<<"\nEnter the number :";
    cin>>n;
    rev = reverse(n);
    cout<<"\nThe reverse of number is :"<<rev;
}
    int reverse(int p)
{
    int d;
    static int r = 0;
    if ( p == 0)
        return (p);
    else
{
```

```
        d = p % 10;
        r = r * 10 + d;
        reverse( p/10);
    }
    return(r);
    }
```

OUTPUT:

```
        Enter the number :
        1234
        The reverse number is :
        4321
```

Q. 1.46. Write a program to calculate the power of an input number using recursion.

Ans.
```
#include<iostream.h>
    int power(int, int);
    void main()
    {
        int n, p, ans;
        cout<<"\n Enter the number:";
        cin>>n;
        cout<<\n Enter the power for number:";
        cin>>p;
        ans = power(n,p);
        cout<<"\n The calculated answer is :"<<ans;
    }
    int power(int n, int p)
    {
        static int r;
        if(p == 0)
        return(1);
    else
    {
        r = r * n;
        power(n, p-1);
    }
    return(r);
    }
```

OUTPUT:

```
        Enter the number:
        5
        Enter the power for number:
        2
        The calculated answer is :
        25
```

Q. 1.47. Write a program to check an input number is prime or not using recursion.

Ans.
```
#include<iostream.h>
int check_prime(int);
void main()
{
    int n, p = 0;
    cout<<"\n Enter the number:";
    cin>>n;
    if(n == 2)
        cout<<"\n Number is Prime";
else if(n < 2)
    cout<<"\n The smallest Prime number is 2";
else
{
    p = check_prime(n);
    if(p)
        cout<<"\n Number is Prime";
    else
        cout<<"\n Number is not Prime";
}
int check_prime(int n)
{
    static int d = 2, status = 1;
    if ( d == n)
        return(0);
    else
{
    if ( n % d == 0)
    {
        status = 0;
        d++;
    }
    else
    {
        d++;
        status = 1;
        check_prime(n);
    }
}
return(status);
}
```

OUTPUT:

```
Enter the number:
5
Number is Prime
Enter the number
6
Number is not Prime
```

Q. 1.48. Write a program to display the series of prime numbers using recursion.

Ans.
```
#include<iostream.h>
void prime(int);
void main()
{
    int range;
    cout<<"\nEnter the Range:
    cin>>range;
    cout<<"\n 2";
    prime(range-1);
}
void prime(int n)
{
    static int a = 3;
    int status = 0;
    if(n != 0)
    {
        for( int i = 2; i<= a-1; i++)
        {
            if( a % i != 0)
                status = 1;
            else
            {
                status = 0;
                break;
            }
        }
        if(status)
        {
            cout<< a<<"\t";
            a++;
            prime(n-1);
        }
        else
        {
            a++;
            prime(n);
        }
    }
}
```

OUTPUT:

```
Enter the Range:
4
2   3   5   7
```

Q. 1.49. Write a program to display greatest common divisor between two number using recursion.

Ans.
```
#include<iostream.h>
int gcd(int,int);
void main()
{
    int a,b,GCD;
    cout<<"\n Enter two numbers :";
    cin>>a>>b;
    GCD = gcd(a,b);
    cout<<"\n G.C.D is :"<<GCD;
}
int gcd(int x, int y)
{
    int r;
    if (y == 0)
        return (x);
    else
    {
        r = x % y;
        return ( gcd(y,r));
    }
}
```

OUTPUT:

```
    Enter two numbers
    5
    8
    G.C.D. is :
    1
```

**Q. 1.50. Write a program to calculate the ^nC_r using recursion.
Where ^nC_r = n!/((n-r)! * r)**

Ans.
```
#include<iostream.h>
int ncr(int,int)
void main()
{
    int n, r, NCR;
    cout<<"\n Enter the values of n and r:";
    cin>>n>>r;
    NCR = ncr(n,r);
    cout<<"\n Answer is :"<<NCR;
}
int ncr(int x, int y)
{
    if(y > x)
{
    cout<<"\n Value of n is less than r, its invalid";
    exit(1);
}
if (y == 0 || y == x)
    return (1);
```

```
else
     return( ncr(x-1, y-1) + ncr(x-1,y));
}
```

OUTPUT:

```
      Enter the values of n and r :
6
2
Answer is :
15
Enter the values of n and r :
2
6
The value of n is less than r, its invalid
```

Q. 1.51. Write a program to find the smallest and greatest number from array of numbers.

Ans.
```
#include<iostream.h>
void main()
{
     int a[10], max, min;
     cout<<"\n Enter the numbers in array:";
     for(int i=0;i<10;i++)
         cin>>a[i];
     cout<<"\n Entered array is:"<<endl;
     for(i=0;i<10;i++)
         cout<<a[i]<<"\t";
     max = a[i];
     min = a[i];
     for(i=0;i<10;i++)
{
     if( max < a[i]);
         max = a[i];
     if( min > a[i]);
         min = a[i];
}
     cout<<"\n Greatest Number is :"<< max;
     cout<<"\n Smallest Number is :"<< min;
}
```

OUTPUT:

```
Enter the numbers of array:
1
23
5
67
3
45
12
89
```

```
34
56
Entered array is :
1   23   5   67   3   45   12   89   34   56
Greatest Number is:
89
Smallest Number is:
1
```

Q. 1.52. Write a program to sort the numbers of array. (Bubble Sort)

Ans.
```
#include<iostream.h>
void main()
{
    int a[10], temp;
    cout<<"\n Enter the numbers in array:";
    for(int i=0;i<10;i++)
        cin>>a[i];
    cout<<"\n Entered array is:"<<endl;
    for(i=0;i<10;i++)
        cout<<a[i]<<"\t";
    for(i=0;i<10;i++)
    {
        for(int j=0;j<10;j++)
    {
        if(a[i] > a[j+1])
    {
        temp = a[j];
        a[j] = a[j+1];
        a[j+1] = temp;
    }
    }
    }
    cout<<"\n Sorteed array is:"<<endl;
    for(i=0;i<10;i++)
    cout<<a[i]<<"\t";
}
```

OUTPUT:

```
Enter the numbers of array:
1
23
5
67
3
45
12
89
34
56
```

```
Entered array is :
1   23  5   67  3   45  12   89   34   56
Sorted array is :
1   3   5   12  23  34  45   56   67   89
```

Q. 1.53. Write a program to calculate the frequency of given number in array of numbers.

Ans.

```
#include<iostream.h>
void main()
{
     int a[10], n, count=0;
     cout<<"\n Enter the numbers in array:";
     for(int i=0;i<10;i++)
          cin>>a[i];
     cout<<"\n Entered array is:"<<endl;
     for(i=0;i<10;i++)
          cout<<a[i]<<"\t";
     cout<<"\n Enter the number to check the frequency:;
     cin>>n;
     for(i=0;i<10;i++)
{
     if(n == a[i])
          count++;
}
     cout<<"\n The frequency of "<< n << "is:"<<count;
}
```

OUTPUT:

```
Enter the numbers of array:
1
23
3
67
3
45
3
89
3
56
Entered array is :
1   23  3   67  3   45   3   89   3   56
Enter the number to check the frequency:
3
the frequency of 3 is : 4
```

Q. 1.54. Write a program to convert an input lower case string in upper case.

Ans.

```
#include<iostream.h>
void main()
{
     char str[20],*s;
```

```
        cout<<"\n Enter the string :";
        cin>>str;
        s=str;
        while(*s!='\0')
        {
        if(*s>=97 && *s<=122)
        *s=*s-32;
        *s++;
    }
    cout<<"\n String is upper case :"<< str;
    }
```

OUTPUT:

```
        Enter the string :
        Computer
        String is in upper case :
        COMPUTER
```

Q. 1.55. Write a program to find out the factorial of given number using function.

Ans.

```
# include <iostream.h>
  int factorial (int n);
  void main ()
        {
              int fact, num;
              cout << "/n Enter the number;"
              cin >> num;
              fact = factori2al (num);
              cout << "/n Factorial of this number = " << fact;
              }
        int factorial (int n)
        {
              f = 1;
              for (i = 1; i < = n; i ++)
                    f = f *i;
                    return (f) ; }
```

❑❑❑

2

OBJECT ORIENTED CONCEPT

CHAPTER OUTLINE

2.1 INTRODUCTION

Object Oriented programming was developed due to the limitations of the traditional programming approaches. The traditional programming approaches, such as Pascal, C, BASIC, Fortran and etc., are basically called *procedural-oriented programming* languages.

Procedural-oriented programming basically emphasis on writing a list of instructions to tell the computer to do something: Get some input, add these numbers, divide by 6, display that output. The programs are dividend into a small sub-program know as a function. Most of functions share global data, if program is too large it is very difficult to identify what data is used by which function. It does not model real world problem, and new enhancement is not easy, whole will be changed or new will be develop.

2.2 BASICS FEATURES OF OOP *[Most Important]*

Object Oriented Programming emphasis is on data rather than function. It divides programs into objects, and they can communicate with each other. It provides the concept to hide the data from the external function and program. New data and functions can be easily added wherever necessary. Some fundamental features of Object-Oriented programming are :

- **Encapsulation:** Encapsulation is a mechanism of binding the member data and member function in a single unit.
- **Data Abstraction:** Data abstraction is mechanism to provide the essential feature without describing the background details.
- **Inheritance:** Inheritance is a mechanism of sharing the member data and member function among the classes.
- **Polymorphism:** Polymorphism is a property to share a single item (or name) in more than one form.

There are several OOP Programming Languages are available, Such as :
• JAVA, • C#, • Simula, • Small • Talk, • Ada, • Eiffel

2.2.1 Encapsulation

In object oriented programming, encapsulation is a basic feature. Encapsulation is a mechanism to place related data and related code in a single unit.

Definition : Encapsulation is a feature to bind the member data and member function in a single unit.

The class and structure is a good example of encapsulation in C++. See an example hoe to bind the data and function in c++ to achieve encapsulation :

Class as Encapsulation :

```
class Employee{
int empcode;
    float salary;
    void getdata(){ ...}
    void putdata(){ ...}
    };
```

Structure as Encapsulation :

```
struct Student{
    int rollno;
    float marks;
    void inputdata(){ ...}
    void outputdata(){ ...}
    };
```

2.2.2 Abstraction

In object oriented programming, Data abstraction is mechanism to provide the essential feature without describing the background details. Means providing functions in class or structure to access data which may be private or public data.

You will learn a private or public in section 2.4.

Definition : Abstraction is a mechanism to provide the essential features without describing the background details.

Abstraction means hide member data or member function of class from outer world. We can achieve abstraction declaring them as a private and providing some member functions for access them in outside. We can access any private member from public members of the same class.

See the structure example :

```
struct Student{
     int rollno;
     float marks;
     void inputdata(){ ...}
     void outputdata(){ ...}
     };
```

The object (variable) of that structure can access rollno and marks easily outside the structure, but in case the data are declared as a private than object cannot able to access them outside the structure. See example :

```
struct Student{
     private:
          int rollno;
          float marks;
     public:
          void inputdata(){ ...}
          void outputdata(){ ...}
          };
```

In this situation we can declare function **inputdata()** and **outputdata()** as public and they can access rollno, and marks easily, either it is private or public.

The member functions which are declared as public can be access easily any other function and out of this structure.

So this is a technique of abstraction which provides functions to access private or public data.

2.2.3 Inheritance

One of the most powerful feature of classes in C++ is that they can be extended through inheritance. The feature of reusability, implements by the *inheritance* in C++ language.

Definition : Inheritance is a mechanism to share the members among the classes.

Inheritance means taking an existing class and adding functionality by deriving a new class from it.

The class you start with, is called the *base class*, and the new class you create, is called the *derived class*.

When you derive a class from another class, the new class gets all the functionality of the base class plus whatever new features you add. You can add data members and functions to the new class, but you cannot remove anything from what the base class offers.

We will discuss the inheritance in details later in chapter 5..

2.2.4 Polymorphism

The Literal mean of Polymorphism (Poly + Morphism) is multiple forms of anything. So in object oriented programming, Polymorphism is a property to share a single item (or name) in more than one form.

Definition : Polymorphism is a mechanism to provide the ability to design multiple forms of a single identifier.

In C++ there are several ways to achieve polymorphism :

- Function overloading,
- Function overriding,
- Operator overloading,
- Virtual functions, etc

Related Questions In Exams

Q.1. Differentiate between procedural abstraction and data abstraction.[Dec-2006]

Ans. Abstraction : Abstraction means hide member data or member function of class from outer world. We can achieve abstraction declaring them as a private and providing some member functions for access them in outside. We can access any private member from public members of the same class.

Procedural Abstraction : When we declare any member function as private in the class, now the function can't be access directly outside the class. It means, procedure hidden from outside the class. This mechanism called *procedure abstraction.*

Data Abstraction : When we declare any member data as private in the class, now the data can't be access directly outside the class. It means, data hidden from outside the class. This mechanism called *data abstraction.*

Example :

```
class Test
{
    private :
    int x;
    void getdata()
    {
        cin >> x;
    }
public:
    void putdata()
    {
        getdata();
        cout << x;
    }
}
```

Members are hidden from outside the class

Member can be access from outside the class, it can be access private member of class

Q.2. Explain Data abstraction with example. *[Dec-2008]*

Ans. Refer Question 1.

Q.3. What are the basic features of OOPs? [May-2009]

Ans. Refer Section 2.2.

Q.4. What are the concept of OOPs? *[May-2008]*

Ans. Refer Section 2.2.

Q.5. What is Object Oriented programming? How is it different from Procedure-Oriented programming? *[Dec-2008]*

C++ follows all Basic OOP features : Encapsulation Abstraction Inheritance Polymorphism.

In What way object oriented paradigm is better than structure programming paradigm? Dec-2009]

Ans. Characteristics of Structural Programming:

1. Large programs are divided into smaller programs known as functions.
2. Most of the functions share global data.

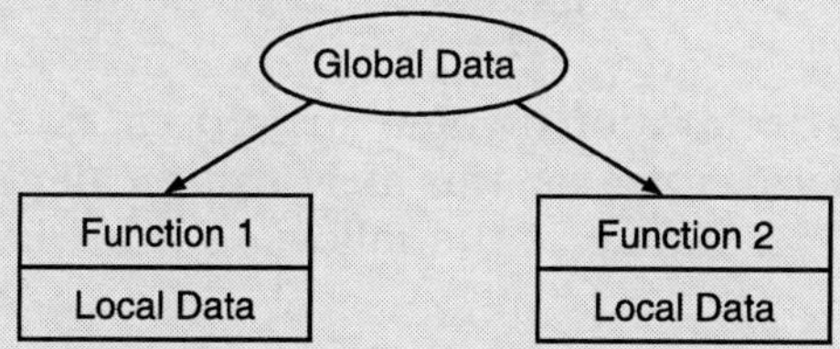

3. Emphasis is on doing things (algorithms).
4. Data move openly around the system from function to functions.
5. Functions transforms data from one form to another.
6. They exhibit "top-down" approach in program design.

Characteristics of object-oriented Programming:

1. It does not allow data to move freely around the system. It ties data more closely to the functions that operate on it.
2. It protects data from accidental modifications from outside functions.
3. It emphasises on data rather than procedures.
4. It decomposes a problem into a number of entities called objects and then builds data and functions around these entities.
5. Data is hidden and cannot be accessed by external functions.
6. Objects may communicate with each other through functions.
7. It follows "button-up" approach in program design.
8. Objects is considered to be partitioned area of complex memory that stores data and set of operations that can access that data.

Q.6. Explain Encapsulation with example. [May-2008]

Ans. Refer Section 2.2.

Q.7. Differentiate between the Public, private and protected access type? [Dec-2007]

What are the access privileges in C++? What is the default access level?

Ans. Refer Section 2.4 , **Default access** Level is private.

2.3 OBJECT AND CLASS

A ***class*** is an expanded concept of a data structure: instead of holding only data, it can hold both data and functions.

An ***object*** is an instantiation of a class. In terms of variables, a class would be the type, and an object would be the variable.

Classes are generally declared using the keyword *class*, with the following format :

```
class class_name
{
    access_modifier_1:
        member1;

    access_modifier_2:
```

```
        member2;
    ...
};
```

Where class_name is a valid identifier for the class. The body of the declaration can contain members, that can be either data or function declarations, and optionally access modifiers.

All is very similar to the declaration on data structures, except that we can now include also functions and members, but also this new thing called *access modifier or access specifiers.* An access specifier is one of the following three keywords: private, public or protected.

For example :

```
class Student{
    private:
        int rollno;
        float marks;
    public:
        void inputdata(){ …}
        void outputdata(){ …}
        };

Student amit;
```

In above example **Student** is a class and **amit** is an object.

2.4 ACCESS MODIFIERS

The modifiers modify the access rights that the members following them acquire :

- **Private** members of a class are accessible only from within other members of the same class or from their *friends.*
- **Protected** members are accessible from members of their same class and from their friends, but also from members of their derived classes.
- Finally, **public** members are accessible from anywhere where the object is visible.

By default, all members of a class declared with the class keyword have private access for all its members. Therefore, any member that is declared before one other class modifiers automatically has private access. For example :

```
class CRectangle
{
    int x, y;           // variable declaration
                        // private by default
public:
    void set_values (int,int);
        int area (void);
    };
```

Declares a class (*i.e.,* a type) called CRectangle. This class contains four members: two data members of type int (member x and member y) with private access (because private is the default access level) and two member functions with public access: set_values() and area(), of which for now we have only included their declaration, not their definition.

Private : *No-body can access except same class members and friend function.*
Protected : *Treated as private but can be access in derived classes.*
Public : *Any where can access in the program.*

Related Questions In Exams

Q.1. Differentiate between the Public, private and protected access type? *[Dec-2007]*
What are the access privileges in C++? What is the default access level? *[Dec-2009]*
Ans. Refer Section 2.4 , Default access Level is private.

2.5 ABSTRACT CLASSES

An abstract class is a class that serves only as a base class from which classes are derived. No object of an abstract base class are created. Thus a base class which have no direct object is called abstract class. A base class that contains at least one pure virtual function is also an abstract base class (see more detail in next chapter in section of virtual function).

Summary

- Object oriented programming emphasis is on data rather than function. Object oriented programming provide the concept to hide the data from the external function and programs. the basic features of object oriented programming are Encapsulation, Abstraction, Inheritance, and Polymorphism.
- Encapsulation is a mechanism of binding the member data and member function in a single unit.
- Data Abstraction is a mechanism to provide the essential feature without describing background details.
- Inheritance is a mechanism of sharing the member data and member function among the classes.
- Polymorphism is a property to share a single item (or name) in more than one form.

Solved Problems

Q. 2.1. What are the main characteristics of structural programming?
Ans. Characteristics of Structural Programming:

1. Large programs are divided into smaller programs known as functions.
2. Most of the functions share global data.

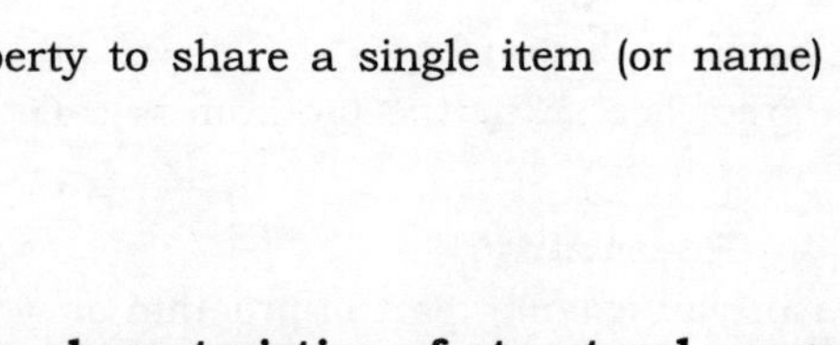

3. Emphasis is on doing things (algorithms).

4. Data move openly around the system from function-to-functions.
5. Functions transforms data from one form to another.
6. They exhibit "top-down" approach in program design.

Q. 2.2. What are the main characteristics of object oriented programming?

Ans. Characteristics of object-oriented Programming:

1. It does not allow data to move freely around the system. It ties data more closely to the functions that operate on it.
2. It protects data from accidental modifications from outside functions.
3. It emphasises on data rather than procedures.
4. It decomposes a problem into a number of entities called objects and then builds data and functions around these entities.
5. Data is hidden and cannot be accessed by external functions.
6. Objects may communicate with each other through functions.
7. It follows "button-up" approach in program design.
8. Objects is considered to be partitioned area of complex memory that stores data and set of operations that can access that data.

Q. 2.3. Given some name of programming languages for structured programming and object oriented programming.

Ans. Structured Programming Languages (Procedural Oriented):

1. C
2. Pascal
3. Fortran
4. Cobol

Object-oriented Programming Languages:

1. C++
2. Java
3. Simula
4. Small talk
5. Ada
6. Eiffel

Q. 2.4. What do you mean by Data hiding?

Ans. Insulation of data from direct access by the program is called data hiding or information hiding.

Q. 2.5. What do you mean by Reusability?

Ans. The reusability means, using an existing part of program or whole program in another new program. We can add additional features to an existing class without modifying. This is possible by deriving a new class from the existing one. The new class will have the combined features of both the classes. The feature of reusability can be achieved by inheritance.

Q. 2.6. What is Dynamic Binding?

Ans. Dynamic binding means that the code associated with a given procedure is not known until the time of the call at run time. It is also known as **late binding**.

Q. 2.7. What do you mean by object-based programming? *(Important)*

Ans. The object oriented programming is the super-set of object-based programming.

Object-based programming is the style of programming that primarily supports encapsulation and object-identity. Major features that are required for object-based programming are :

- Data abstraction

- Data hiding and access mechanisms
- Automatic initialization and clear-up of objects
- Operator over-loading.

Languages that support programming with objects are said to be object-based programming languages.

Object oriented programming incorporates all of object-based programming features along with two additional features-inheritance and dynamic binding.

Object Orient Programming ⇒

- Object-based features
  +
- Inheritance
  +
- Dynamic binding.

Exercise

Q. 2.1. What do you mean by Object Oriented Programming ?

Q. 2.2. What are the basic features of Object Oriented Programming ?

Q. 2.3. What is Object ?

Q. 2.4. What is Class ?

Q. 2.5. What is difference between Object and Class ?

Q. 2.6. What are the access modifiers are available in C++ ?

Q. 2.7. What is abstract class ?

Q. 2.8. Write difference between

(a) Object and classes.

(b) Data abstraction and encapsulation.

(c) Inheritance and Polymorphism.

(d) Structural programming and object oriented programming.

Q. 2.9. Explain data abstraction and Data hiding.

❑❑❑

3

CLASSES AND DATA ABSTRACTION

CHAPTER OUTLINE

- ➢ **Structure**
  - — **Declaration**
  - — **Accessing Members**
- ➢ **Class**
  - — **Declaration**
  - — **Accessing Members**
  - — **Controlling Functions**
- ➢ **Constructors**
  - — **Default Constructor**
  - — **Parameterised Constructor**
  - — **Copy Constructor**
  - — **Destructors**
- ➢ **Class Members**
  - — **Object Member**
  - — **Constant Members**
  - — **Static Members**
- ➢ **Inline Function**
- ➢ **Friend Function**
- ➢ **Friend Classes**
- ➢ **Dynamic Memory Allocation**
  - — **Operators new and new[]**
  - — **Operator delete and delete[]**
- ➢ **This Pointer**
- ➢ **Container Classes**
- ➢ **Proxy Classes**
- ➢ **Function Overloading**
- **Summary**

3.1 STRUCTURE

A structure is a collection of one or more variables, possibly of different types, grouped together under a single name for convenient handling. (Structures are called "*records*" in some languages, notably Pascal.) Structures help to organise complicated data, particularly in large programs, because they permit a group of related variables to be treated as a unit instead of as separate entities.

3.1.1 Declaration

The syntax for declaration of a structure is as follows :

Syntax 1

```
struct  1
{
    <data members>;
    <function members>;
    <other members>;
}variable;
```

Example for Syntax 1 :

```
struct
{
    int empid;
    char name [51];
    int basic;
} emp;
```

Syntax 2

```
struct <structname>
{
    <data members>;
    <function members>;
    <other members>;
};

struct <structname> variable;
```

Example for Syntax 2 :

```
struct <structname>
{
    int empid;
    char name'51];
    int basic;
};

Struct employee emp;
```

Syntax 3

```
typedef struct
{
    <data members>;
    <function members>;
    <other members>;
}datatypename;

datatypename variable;
```

Example for Syntax 3 :

```
typedef struct
{
    int empid;
    char name[51];
    int basic;
} employee;

employee emp;
```

3.1.2 Accessing Members

To access members of a structure use **"." (dot)** operator see example :

```
struct employee
{
    int empid;
    char name[51];
    int basic;
};
struct employee emp;
emp.empid=34;
strcpy(emp.name,"Ramesh");
emp.basic=6000;
```

Here are some simple example to understand how to use stuructures in the C++ programs.

Example 3.1 **Display the bio-data using structure**

```
#include<iostream.h>

struct bio
{
    char name[20];
    int age;
    float sal;
    char address[20];
};

void main()
{
struct bio info={ “Rajkumar”, 28, 10000.00, “Haridwar”};

cout<<“Name       :    “<<info.name<<endl;
cout<<“Age        :    “<<info.age<<endl;
cout<<“Salary     :    “<<info.sal<<endl;
cout<<“Address    :    “<<info.address<<endl;
}
```

OUTPUT

```
Name    :    Rajkumar
Age     :    28
Salary  :    10000.00
Address :    Haridwar
```

Structure in C contains data only but in C++ Structure may have data and function both.

Example 3.2 **Read and display the bio-data using structure**

```
#include<iostream.h>
struct bio
{
    char name[20];
    int age;
    float sal;
    char address[20];
};

int main()
{
struct bio info;
cout<<“Enter ur bio-data : name, age, sal, address”<<endl;
cin>>info.name>>info.age>>info.sal>>info.address;
cout<<“Given Biodata :-“<<endl;
```

```
cout<<"Name: "<<info.name<<endl;
cout<<"Age: "<<info.age<<endl;
cout<<"Salary: "<<info.sal<<endl;
cout<<"Address: "<<info.address<<endl;
}
```

OUTPUT

```
Enter ur bio data    :   name, age, sal, address
Rajkumar
28
10000.00
Haridwar
Given Boidata:-
Name     :    Rajkumar
Age      :    28
Salary   :    10000.00
Address  :    Haridwar
```

3.1.3 Structure with functions

Unlike C, we can use functions with the structures, it means we can define the structure which can have functions too. Here you can see in the sample code we define a function in the structure definition.

```
struct demo
    {
        int x,y;
        void input()
        {
            cont<<"\n Enter value of x & y";
            cin>>x>>y;
        }
        void output()
        {
            cont<<"\n value of x:"<<x;
            cont<<"\n value of y:"<<y;
        }
    };
```

In the above code you can see that the method (function) **input()** and **output()** can read and display the values of x and y from/to console.

Now we going to show that, we can also access the member functions using dot "." operator, like a data members.

```
demo d;
d.input();
d.output();
```

The method **input()** and **output()** also helps to reduce the length of code and we can invoke these functions for any object of **demo** structure. Let see an Example :

Example 3.3 **Structure with function**

```
#include<iostream.h>
#include<conio.h>

struct Demo
{
    int x,y;
    void input()
    {
        cout<<"\n Enter value of x & y:";
        cin >> x >> y;
    }
    void output()
    {
        cout<<"\n value of x = " << x;
        cout<<"\n value of y = " << y;
    }
};

int main()
{
    Demo d1, d2;
    d1.input();
    d2.input();
    d1.output();
    d2.output();
    return 0;
}
```

OUTPUT

```
Enter value of x & y;
10 20
Enter value of x & y;
30 40
Value of x = 10
Value of y = 20
Value of x = 30
Value of y = 40
```

3.2 CLASS

So far you've used simple variables and structures to hold data and functions to process the data. C++ *classes* allow you to combine data and the functions that use it.

A ***class*** is an expanded concept of a structure: instead of holding only data, it can hold both data and functions.

An ***object*** is an instantiation of a class. In terms of variables, a class would be the type, and an object would be the variable.

3.2.1 Declaration

Classes are generally declared using the keyword *class*, with the following format :

```
class class_name
{
access_modifier_1:
    member1;
access_modifier_2:
    member2;
...
};
```

Where class_name is a valid identifier for the class. The body of the declaration can contain members, that can be either data or function declarations, and optionally access modifiers.

All is very similar to the declaration on data structures, except that we can now include also functions and members, but also this new thing called *access modifier or access specifiers*. An access specifier is one of the following three keywords: private, public or protected.

For example :

```
class Student{
    private:
        int rollno;
        float marks;
    public:
        void inputdata(){ …}
        void outputdata(){ …}
        };
```

3.2.2 Accessing Members

To access the members of class you have to create objects, which also known as class variables.

Like a declaration of a variables of any built-in data types, you can create object, which is also known as variable of class. Once a class has been declared, we can create variables of that type by using the class name. For example :

```
Student amit;
```

In this example, **Student** is the class name (*i.e.,* the type), whereas **amit** is an object of type *item.*

Example 3.4 **Accessing the Class members with single object**

```
#include <iostream.h>
class demo
{
    private:
        int x,y;
    public:
        int a,b;
        void getdata();
        void putdata();
};

void demo :: getdata()
{
        cout<<"\n Enter the values of x and y:"<<endl;
        cin>> x >> y;
        cout<<"\n Enter the values of a and b:"<<endl;
        cin>>a>>b;
}

void demo :: putdata()
{
        cout<<"\n The value of x is :" << x << " and y is : "<< y;
        cout<<"\n The value of a is :" << a << " and b is : "<< b;
}

void main()
{
        demo D;
        D.getdata();

        cout<<  "\n The Entered Values are:";
        D.putdata();
        D.a = 10;          //public data can be access in
        D.b = 20;          //main or non member function of class

        cout<<  "\n The Values after the moderation are:";
        D.putdata();
        }
```

OUTPUT

```
Enter the values of x and y :
23
79
Enter the values of a and b :
200
100
```

```
The Entered Values are:
The value of x is : 23 and y is : 79
The value of a is : 200 and b is : 100

The Values after the moderation are:
The value of x is : 23 and y is : 79
The value of a is : 10 and b is : 20
```

This program shows that, you can access the public variables outside of the member function of class using this statements :

```
D.a = 10; // public data can be access in
D.b = 20; // non-member functions of class
```

But you can not access the private variable like this in out side the class member function :

```
D.x = 10; // illegal, can't access the private data in
D.y = 20; // non-member functions of class
```

3.2.3 Controlling Functions

The member functions of class can define inside and out side the classes. There are some examples to explain the methods for above :

Inside definition :

It is possible that, the function can be define within the class at the time of class declaration. See example :

Example 3.5 **Inside declaration of member function**

```
#include <iostream.h>
class demo
{
    private:
        int x,y;
    public:
        int a,b;
void getdata()
{
        cout<<"\n Enter the values of x and y:"<<endl;
        cin>> x >> y;
        cout<<"\n Enter the values of a and b:"<<endl;
        cin>>a>>b;
}

void putdata()
{
        cout<<"\n The value of x is :" << x << " and y is : "<< y;
        cout<<"\n The value of a is :" << a << " and b is : "<< b;
}
}; //end of class

void main()
```

```
{
        demo D;

        D.getdata();

        cout<<   "\n The Entered Values are:";
        D.putdata();

}
```

OUTPUT

```
Enter the values of x and y :
23
79
Enter the values of a and b :
200
100
The Entered Values are:
The value of x is    :    23 and y is        :      79
The value of a is    :    200 and b is       :      100
```

> ✍ **Note :**
> Using definition within the class, the function is declared and defined entirely within the class declaration. The function is an inline function by virtue of the fact that it is contained within the class declaration. The inline keyword is not required.

Outside definition :

You can also define the function outside the class but you have to remember following instructions :

(*i*) You have to declare function inside the class

(*ii*) You have to use scope resolution operator in the signature of function to define member function outside the class as following syntax :

<return type> class_name :: function_name (<arguments>) {... }

See example :

Example 3.6 **Outside declaration of member function**

```
#include <iostream.h>
class demo
{
    private:
        int x,y;
    public:
        int a,b;
        void getdata();
        void putdata();
};

void demo :: getdata()
{
        cout<<"\n Enter the values of x and y:"<<endl;
```

```
        cin>> x >> y;
        cout<<"\n Enter the values of a and b:"<<endl;
        cin>>a>>b;
}

void demo :: putdata()
{
        cout<<"\n The value of x is :" << x << " and y is : "<< y;
        cout<<"\n The value of a is :" << a << " and b is : "<< b;
}

void main()
{
        demo D;

        D.getdata();

        cout<<   "\n The Entered Values are:";
        D.putdata();

        }
```

OUTPUT

```
Enter the values of x and y :
23
79
Enter the values of a and b :
200
100
The Entered Values are:
The value of x is    :    23 and y is  :   79
The value of a is    :    200 and b is :   100
```

Now see how to use private and public member function in the program.

Example 3.7 **Demonstration of private and public functions**

```
#include <iostream.h>
class demo
{
    private:
        int x,y;
        int a,b;
        void getdata();    // defined as a private function
    public:
        void putdata();
};

void demo :: getdata()
{
        cout<<"\n Enter the values of x and y:"<<endl;
```

```
        cin>> x >> y;
        cout<<"\n Enter the values of a and b:"<<endl;
        cin>>a>>b;
}

void demo :: putdata()
{
        getdata();          // you must call the private function
                            // in the member function of class

cout<<"\n The value of x is :" << x << " and y is : "<< y;
        cout<<"\n The value of a is :" << a << " and b is : "<< b;

}

void main()
{
        demo D;

        // D.getdata();     illegal, you cant access private function
                            in non-member function of class

        D.putdata();        // it will call the getdata() function

}
```

OUTPUT

```
Enter the values of x and y :
23
79
Enter the values of a and b :
20
10

The value of x is    :   23 and y is  :  79
The value of a is    :   20 and b is  :  10
```

This program shows that, if you have defined any member function in private block of class, then you must call this function in the public member function of that class, you cant access outside of the member function of class using this statements :

```
D.getdata();     // illegal, cant access the private function
                 // in non-member functions of class
```

Related Questions In Exams

Q.1. How we can access member of a structure? *[May-2008]*

Ans. Refer Section 3.1.1 and 3.1.2.

Q.2. How does C++ structure differ from a C++ class? *[May-2009]*

What are the differences between C++ structure and C++ class? *[Dec-2009]*

Ans.

Structure	***Class***
1. Uses key word ***struct*** to declare.	1. Uses keyword ***class*** to declare.
2. By default member data are public.	2. By default member data are private.
3. By default member functions are public.	3. By default member functions are private.

4. Syntax of structure :
struct <structure name>
{
<access modifier>:
data members;
function members;
};
	4. Syntax of structure :
	class <class name>
	{
	<access modifier>:
	data members;
	function members;
	};

5. Syntax of declaration of object of : struct <struct name> <object name1>; or <struct name> <object name1>;	5. Syntax of declaration of object of : <class name> <object name1>;

3.3 CONSTRUCTORS

Constructors are distinguished from all other member functions by having the same name as the class they belong to. They used to create, initialise, and copy the objects of a class. When an object of that class is created or is being copied, the appropriate constructor is called *implicitly*. The main characteristics of constructors are as follows :

Characteristics of Constructors

- Constructors do not have return type or value. (not even void).
- The compiler automatically calls constructors defining objects.
- They have same name as class name.
- Constructors always exist in the public block of class.
- Constructors cannot be inherited, though a derived class can call the base class constructors. (we will discuss in later chapters)
- Constructors, like most C++ functions, can have default arguments or use member initialisation lists.
- Constructors cannot virtual. (See virtual destructors)
- You cant take addresses of constructors.
- You cannot call constructors the way you call a normal function.
- Constructors can make implicit calls to operator new if allocation is required for an object.
- An object with a constructor cannot be used as a member of a union.

Constructor definition :

Constructors are must de define as a member function, having the same name as the class they belong to. For example :

```
class Demo
{
```

```
    public:
        Demo();  // constructor of class Demo
};
```

Basically constructors are defined for the initialise the member data of the class and creation of objects. You can initialise the variable of class in which it is defined. For example :

```
class Demo
{
private:
    int a, b;
public:
    Demo()  // definition of constructor
        {
        a = 0;
        b = 0;
        }
};
```

Types of Constructors

There are three types of constructors are available in C++.

1. default constructor
2. parameterised constructor
3. copy constructor
4. dynamic constructor

3.3.1 Default Constructor

The default constructor for class Demo is one that takes no arguments. If no user-defined constructors exist for a class, C++ compiler generates a default constructor. On a declaration such as Demo D, the default constructor creates the object D. For example :

```
class Demo
{
private:
    int a, b;
public:
        Demo()
        {
            a = 0;
            b = 0;
            }
};

void main()
{
    Demo D;
    }
```

Default constructor with definition

```
class Demo
{
private:
        int a, b;
};

void main()
{
        Demo D;
}
```

Default constructor without definition

Here we have given some examples of programs to understand the working with constructors.

Example 3.8 **Default constructor**

```
#include <iostream.h>
class Demo
{
    private:
        int x,y;
    public:
        Demo();
        void getdata();
        void putdata();
};

Demo :: Demo()
{
    x = 0;
    y = 0;
}
void Demo :: getdata()
{
    cout<<"\n Enter the values of x and y:"<<endl;
    cin>> x >> y;

}

void Demo :: putdata()
{
    cout<<"\n The value of x is :" << x << " and y is : "<< y;
}

void main()
{
    Demo D;

    D.putdata();       // it will display the initial values of member
                       data of Demo
                       // which initialized by default constructor
    D.getdata();

    D.putdata();
}
```

OUTPUT

```
The value of x is : 0 and y is : 0

Enter the values of x and y :
23
```

```
79

The value of x is : 23 and y is : 79
```

Example 3.9 **Default constructor to print message**

```
#include <iostream.h>
class Demo
{
    public:
        Demo();
        void display();
};

Demo :: Demo()
{
        cout<<"\n I am default constructor";
}

void Demo :: display()
{
        cout<<"\n I am a display function";
}

void main()
{
    Demo D;

    D.display();

    Demo D1, D2;
    D1.display()
}
```

OUTPUT

```
I am default constructor ⟶  output of default constructor of
                             object D
I am display function.   ⟶  output of display function of object D

I am default constructor ⟶  output of default constructor of
                             object D1
I am default constructor ⟶  output of default constructor of
                             object D2
I am display function.   ⟶  output of display function of object
                             D1
```

3.3.2 Parameterised Constructor

A parameterised constructor for class Demo is one that can be called with arguments of any type except Demo. User can define more than one constructor in a single class, but all constructors must have different signature (means number of arguments or types of argument does not match with other constructor in the same class). This property is called *constructor overloading* (we will discuss in later chapters).

```
class Demo
{
    private:
        int a, b, c;
    public:
            Demo(int x, int y, int z)
            {
                a = x;
                b = y;
                c = z;
                }
    };

void main()
{
            Demo D1(3, 7, 10); // parameterized constructor
}
```

Example 3.10 **Parameterised constructor**

```
#include <iostream.h>
class Demo
{
private:
    int x,y;
public:
    Demo()                  //default constructor
    {
        x = 0;
        y = 0;
    }
    Demo(int a, int b)      //parameterised constructor with two
                            arguments
    {
        x = a;
        y = b;
    }

    Demo(int a)             //parameterised constructor with single
                            arguments
    {
        x = a;
        y = x + 10;
    }

        void putdata();
};
```

```
void Demo :: putdata()
{
    cout<<"\n The value of x is :" << x << " and y is : "<< y;
}

void main()
{
    Demo D;              // call default constructor

    Demo D1(10, 15);     //parameterised constructor with two
                         argument

    Demo D2(5);          // parameterised constructor with single
                         argument
    D.putdata();
    D1.putdata();
    D2.putdata();
    int a,b;
    cout<< "\nEnter the values of
    a and b :"<<endl;
    cin>> a >> b;
    Demo S(a,b);
    S.putdata();
}
```

Constructors invoked automatically when object is created and used for : Member Initialisation, Memory Allocation, Object Creation.

OUTPUT

```
The value of x is : 0 and y is : 0      ⟶   output of putdata() of
                                             object D
The value of x is : 10 and y is : 15    ⟶   output of putdata() of
                                             object D1
The value of x is : 5 and y is : 15     ⟶   output of putdata() of
                                             object D2
  Enter the values of a and b:
  3
  4
The value of x is : 3 and y is : 4      ⟶   output of putdata() of
                                             object S
```

3.3.3 Copy Constructor

Copy constructors are invoked when initialising a class object, typically when you declare with initialisation by another class object. A copy constructor for class Demo is one that can be called with a single argument of type Demo, as follows :

```
class Demo
{
private:
    int a, b;
public:
        Demo(int x, int y)
        {
```

```
                a = x;
                b = y;
            }
            Demo(Demo &D)       // definition of copy constructor
            {
                a = D.a;
                b = D.b;
            }
    };
void main()
{
            Demo D1(3, 7);      // parameterized constructor
            Demo D2 = D1;       // copy constructor can be call with
                                " = " operator
            Demo D3 (D1);       // copy constructor can be call with
                                "( )" operator
}
```

Example 3.11 Copy constructor

```
#include <iostream.h>
class Demo
{
    private:
        int x,y;
    public:
        Demo();                     //default constructor
        Demo(int a, int b);         //parameterized constructor with two
                                    arguments
        Demo(int a);                //parameterized constructor with
                                    single arguments
        Demo(Demo &D);              //copy constructor
        void putdata();
};

Demo :: Demo()                      //default constructor
{
        x = 0;
        y = 0;
}

Demo :: Demo(int a, int b)          //parameterized constructor with two
                                    arguments
{
        x = a;
        y = b;
}

Demo :: Demo(int a)                 //parameterised constructor with
```

```
                                        single arguments
{
        x = a;
        y = x + 10;
}

Demo :: Demo(Demo &D)
{
        x = D.x;
        y = D.y;
}

void Demo :: putdata()
{
        cout<<"\n The value of x is :" << x << " and y is : "<< y;
}

void main()
{
        Demo D;                 // call default constructor

        Demo D1(10, 15);        //parameterised constructor with two
                                argument

        Demo D2(5);             // parameterised constructor with
                                single argument

        Demo D3 = D;            //copy constructor invoked with
                                object D as argument

        Demo D4 (D1);           //copy constructor invoked with
                                object D1 as argument

        Demo D5 = D2;           //copy constructor invoked with
                                object D2 as argument

        D3.putdata();

        D4.putdata();

        D5.putdata();

}
```

OUTPUT

```
The value of x is : 0 and y is : 0    ⟶   output of putdata() of
                                           object D3
The value of x is : 10 and y is : 15  ⟶   output of putdata() of
                                           object D4
```

```
The value of x is : 5 and y is : 15    ⟶  output of putdata() of
                                           object D5
```

3.3.4 Dynamic Constructor

Basically constructors are defined for the initialise the member data of the class and creation of objects. When any constructor which uses dynamic allocation is called *dynamic constructor*. It means when any constructor allocate the memory at runtime using new keyword is called dynamic constructor. This is also used to allocate the right size of memory for each object. It can save the memory where same class objects are not of same size. See an example to understand the dynamic constructor and new operator :

Example 3.12 **Dynamic constructor**

```
#include <iostream.h>
#include <string.h>
class Demo
{
    private:
        char *string;
        int length;
    public:
        Demo()                  //default constructor
        {
            length = 0;
            string = new char[length+1];   //dynamic allocation
            strcpy(string,"");
        }

        Demo(char *s)           //parameterised constructor
        {
            length = strlen(s);
            string = new char[length+1];   //dynamic allocation
            strcpy(string,s);
        }

        void putdata();
};

void Demo :: putdata()
{
    cout<<"\n String : "<<string;
}

void main()
{
    Demo D;                         // default constructor call
    Demo D1("vaibhav");             //parameterised constructor
    Demo D2("Nupur Bhatnagar");
    D.putdata();
    D1.putdata();
```

```
    D2.putdata();
}
```

OUTPUT

```
String :
String : vaibhav
String : Nupur Bhatnagar
```

You can see that first D object of demo class allocated only for one byte, second object D1 allocated for 8 bytes, third object D2 allocated for 16 bytes. It means all objects having different size in the memory. That the benefit of dynamic constructor.

3.3.5 Destructors

Whenever you declare a constructor, you'll also want to declare a destructor. Just as constructors create and initialise objects of your class, destructors clean up after your object and free any memory you might have allocated. A destructor always has the name of the class, preceded by a tilde (~). Destructors take no arguments and have no return value.

Characteristics of Destructor :

- Destructors do not have return type or value. (not even **void**).
- The compiler automatically calls destructors when destroying objects.
- Destructors have same name as class name, preceded with a tilde symbol(~).
- Destructors always exist in the public block of class.
- Destructors cannot be inherited, though a derived class.
- Destructors can be virtual, but constructors cannot. (See Virtual destructors.)
- You cant take their addresses.
- Destructors can be called if you use their fully qualified name.
- Destructors can make implicit calls to operator delete if required for an object.
- An object with a constructor or destructor cannot be used as a member of a union.

Destructor Definition

Destructors are must de define as a member function, having the same name as the class they belong to, preceded with tiled (~) symbol. For example :

```
class Demo
{
    public:
        Demo();         // constructor of class Demo
        ~Demo(){ }      //destructor of calss Demo
};
```

If a destructor isn't explicitly defined for a class, the compiler generates one.

Invoking Destructors

A destructor is called implicitly when a variable goes out of its declared scope. Destructors for local variables are called when the block they are declared in is no longer active. In the case of global variables, destructors are called as part of the exit procedure after the main function.

When pointers to objects go out of scope, a destructor is not implicitly called. This means that the delete operator must be called to destroy such an object.

Destructors are called in the exact opposite order from which their corresponding constructors were called. For example :

Example 3.13 **Invoking order of destructor**

```
#include <iostream.h>
class Demo
{
    public:
        Demo();
        void display();
        ~Demo()
{
        cout<<"\n Bye - Bye";
}
};

Demo :: Demo()
{
        cout<<"\n I am default constructor";
}

void Demo :: display()
{
        cout<<"\n I am a display function";
}

void main()
{
        Demo D, D1, D2;
        D1.display();
}
```

OUTPUT

```
I am default constructor    ⟶   output of default constructor of
                                object D
I am default constructor    ⟶   output of default constructor of
                                object D1
I am default constructor    ⟶   output of default constructor of
                                object D2

I am display function.      ⟶   output of display function of object
                                D1

Bye-Bye                     ⟶   output of destructor of object D2
Bye-Bye                     ⟶   output of destructor of object D1
Bye-Bye                     ⟶   output of destructor of object D
```

A destructor can also be invoked explicitly in one of two ways: indirectly through a call to delete, or directly by using the destructors fully qualified name. You can use delete to destroy objects that have been allocated using new. Explicit calls to the destructor are necessary only for objects allocated a specific address through calls to new.

Related Questions In Exams

Q.1. What is constructor? Explain the use of dynamic constructor, using suitable example. *[Dec-2006]*

What is constructor? Explain do you mean by dynamic initialisation of objects?Give example. *[May-2009]*

Ans. Refer section 3.3 to 3.3.4.

Q.2. What is copy constructor? Why it is required? *[May-2008]*

What do you mean by copy constructor? Explain with help of example. *[Dec-2008]*

What is copy constructor and when are they called? Give a suitable example. *[Dec-2009]*

Ans. Refer Section 3.3.3.

Q.3. What is destructor? Where it has been used? Give example. *[Dec-2007]*

Ans. Refer Section 3.3.5.

Q.4. Differentiate between constructor and destructors. [Dec-2006]

Ans.

Constructor	*Destructor*
1. Has same name of class.	1. Has same name of class with tiled (~) symbol.
2. Automatically invoked when object created.	2. Automatically invoked when scope is goes end.
3. Used to initialise the member data of class.	3. Used close some necessary files.
4. Used to allocate the memory for object.	4. Release the memory occupied by the objects.
5. Invoked in same order of creation of objects.	5. Invoked in descending order of object creation.

3.4 CLASS MEMBERS

As you saw in above examples classes contain member data and member functions. The members may be public, private or protected. Member data may be of any data type, array type or pointer type. Member functions may be of any type, such as with return or without return value, with arguments or without arguments.

But class may have some user defined data types such as objects, or data may be constant or static type. Let me explain.

3.4.1 Object Member

A data member of a class may be of a user-defined type, that is, an object of another class. See Example :

```
class Point
{
    private:
        int x,y;
};
```

```
class Rectangle
{
    private:
    Point topLeft;
    Point botRight;
};
```

In above example, a **Rectangle** class may be defined using two **Point** data members which represent the top-left and bottom-right corners of the rectangle.

3.4.2. Constant Members

A class data member may defined as constant. For example :

```
class Image
{
const int width;
const int height;
//...
};
```

However, data member constants cannot be initialised using the same syntax as for other constants :

```
class Image
{
const int width = 256; // illegal initializer!
const int height = 168; // illegal initializer!
//...
};
```

The correct way to initialize a data member constant is through a member initialisation list :

```
class Image
{
public:
Image (const int w, const int h);
private:
const int width;
const int height;
//...
};
Image::Image (const int w, const int h) : width(w), height(h)
{
//...
}
```

As one would expect, no member function is allowed to assign to a constant data member.

A constant data member is not appropriate for defining the dimension of an array data member. For example, in

```
class Set
{
public:
Set (void) : maxCard(10) { card = 0; }
//...
private:
const maxCard;
int elems[maxCard]; // illegal!
int card;
};
```

The array elems will be rejected by the compiler for not having a constant dimension. The reason for this being that maxCard is not bound to a value during compilation, but when the program is run and the constructor is invoked. Member functions may also be defined as constant. This is used to specify which member functions of a class may be invoked for a constant object. For example,

```
class Set
{
public:
Set (void) { card = 0; }
Bool Member (const int) const;
void AddElem (const int);
//...
};
Bool Set::Member (const int elem) const
{
//...
}
```

Above example defines Member as a constant member function. To do so, the keyword const is inserted after the function header, both inside the class and in the function definition.

A constant object can only be modified by the constant member functions of the class :

```
const Set s;
s.AddElem(10);    // illegal: AddElem not a const member
s.Member(10);     // ok
```

Given that a constant member function is allowed to be invoked for constant objects, it would be illegal for it to attempt to modify any of the class data members.

Constructors and destructors need never be defined as constant members, since they have permission to operate on constant objects. They are also exempted from the above rule and can assign to a data member of a constant object, unless the data member is itself a constant.

3.4.3. Static Members

A data member of a class can be defined to be static. This ensures that there will be exactly one copy of the member, shared by all objects of the class. For example, consider a Number class which represents windows on a bitmap display :

```
class Number
{
static int counter; //
public:
void increment(){ counter++;}
//...
};
```

Here, no matter how many objects of type Number are defined, there will be only one instance of counter. Like other static variables, a static data member is by default initialised to 0. It can be initialised to an arbitrary value in the same scope where the member function definitions appear :

```
Number Number::counter = 100;
```

The alternative is to make such variables global, but this is exactly what static members are intended to avoid; by including the variable in a class, we can ensure that it will be inaccessible to anything outside the class.

Member functions can also be defined to be static. Semantically, a static member function is like a global function which is a friend of the class, but inaccessible outside the class. It does not receive an implicit argument and hence cannot refer to this. Static member functions are useful for defining call-back routines whose parameter lists are pre-determined and outside the control of the programmer.

For example, the Window class might use a call-back function for re-painting exposed areas of the window :

```
class Window {
//...
static void PaintProc (Event *event); // call-back
};
```

Because static members are shared and do not rely on the this pointer, they are best referred to using the *class::member* syntax. For example, counter and PaintProc would be referred to as Number::counter and Window::PaintProc.

Public static members can be referred to using this syntax by no-nmember functions (*e.g.*, global functions).

Example 3.14 **Static Data Member**

```
#include<iostream.h>
#include<conio.h>
class test
{
public:
    static int x;                    // static member data
    void show()
    {
        cout<<x++<<endl;
    }
};
int test::x=100;
int main()
{
```

```
        clrscr();
        test t[4];
        for(int i=0;i<4;i++)
        {
            t[i].show();
        }
        return 0;
    }
```

OUTPUT

```
    100
    101
    102
    103
```

Example 3.15 **Static Member Function**

```
    #include<iostream.h>
    #include<conio.h>
    class test
    {
    public:
        static void show()          // static member function
        {
            cout<<"Welcome"<<endl;
        }
    };
    int main()
    {
        clrscr();
        test ::show();
        return 0;
    }
```

OUTPUT

```
    Welcome
```

In above example no need of object of class test the show() function can be execute using class name using following statement.

```
        test ::show();
```

3.5 INLINE FUNCTION

Every time the calling of function takes a lot of extra time in executing a series of instructions for task such as jumping to the function, saving registers, pushing arguments into stack, and returning to the calling **function**. When a function is small, a substantial percentage of execution time may be spent in such overheads. One solution to this problem is to use of macros instead of functions, but the drawback of macro is that they are not really functions and therefore, the error checking does not occur during compilation.

Thus C++ provides a concept of Inline functions, which works like as macros in C language. It means, Inline functions replaced by its code wherever they called. There are two ways of creating the inline functions.

- Using inline keyword
- Using definition within class

Using inline keyword we can create functions as inline. The syntax is given :

```
inline return-type function-name(arguments)
{
    function-body
}
```

The normal function and member function both can be defined as inline function. There an example for a normal function as inline :

```
inline int sum  (int a, int b, int c)
{
    return (a + b + c);
}
```

And whenever the member functions of class need to be create as inline. Then declare function within the body of the class and the function definition (the function itself) is defined outside the class declaration using the inline keyword must be proceeded by class name and the scope resolution operator. For example :

```
class Demo
    {
        public:
        inline int add (int a, int b);    //function declaration
        };

inline int Demo : : add(int a, int b)    //function definition
        {
            return (a + b);
        }
```

Using definition within the class, the function is declared and defined entirely within the class declaration. The function is an inline function by virtue of the fact that it is contained within the class declaration. The inline keyword is not required. For example :

```
class Demo
    {
    public:
        int Demo::add(int a, int b)  //inline function by default
        {
            return (a + b);
        }
    };
```

3.6. FRIEND FUNCTION

If we want to declare an external (non-member) function as friend of a class, thus allowing this function to have access to the private and protected members of this class, we do it by declaring a prototype of this external function within the class, and preceding it with the keyword friend. The syntax of declaring friend function is :

```
class Class-name
{
public:
friend return-type function-name (arguments)
};
```

Here are given some example to understand the concept of friend function.

Example 3.16 **A simple friend function**

```
#include <iostream.h>
class Demo
{
    private:
        int x,y;
    public:
        Demo();
        void getdata();
        void putdata();
        friend int add();
};

Demo :: Demo()
{
        x = 0;
        y = 0;
}
void Demo :: getdata()
{
        cout<<"\n Enter the values of x and y:"<<endl;
        cin>> x >> y;
}

void Demo :: putdata()
{
        cout<<"\n The value of x is :" << x << " and y is : "<< y;
}

int add()
{
        Demo D;

        D.getdata();
        D.putdata();

        int sum = D.x + D.y;    // addition of private data in friend
                                function return sum;
}
```

```
void main()
{
        int val;

        val = add();            // it will return the sum of private
                                data of Demo class.

        cout<<"\n Sum of private data of class = "<<val ;

}
```

OUTPUT

```
Enter the values of x and y :
2
7

The value of x is : 2 and y is : 7
Sum of private data of class = 9
```

Friend Function can access the private members of class using objects.

In the above example the function add is declared as a friend function of class Demo, which accessed the private data of class, add them and return value to the main function where it is called.

Example 3.17 **Friend function**

```
#include <iostream.h>

class Rectangle
{
    int width, height;
public:

    void set_values (int, int);
    int area () {return (width * height);}

    friend Rectangle duplicate (Rectangle);
};

void Rectangle::set_values (int a, int b)
{

        width = a;
        height = b;
}

Rectangle duplicate (Rectangle B)
{
        Rectangle A;
        A.width = B.width*2;
        A.height = B.height*2;
        return (A);
```

```
}

int main ()
{
        Rectangle R, R1;

        R.set_values (2,3);
        R1 = duplicate (R);
        cout << R1.area();
        return 0;
}
```

OUTPUT

```
24
```

The duplicate function is a friend of Rectangle. From within that function we have been able to access the members width and height of different objects of type Rectangle, which are private members. Notice that neither in the declaration of duplicate() nor in its later use in main() have we considered duplicate a member of class Rectangle. It isn't. It simply has access to its private and protected members without being a member.

The friend functions can serve, for example, to conduct operations between two different classes. Generally, the use of friend functions is out of an object oriented programming methodology, so whenever possible it is better to use members of the same class to perform operations with them. Such as in the previous example, it would have been shorter to integrate duplicate() within the class Rectangle.

Example 3.18 **Friend as a bridge b/w classes**

```
#include<iostream.h>
class B;                // pre declaration of class B before using
class A
{
    int a, b;
public:
    A(int m, int n)
    {
        a = m;
        b = n;
    }

friend add( A, B);
};

class B
{
    int x, y;
public:
    B(int m, int n)
    {
        x = m;
        y = n;
```

```
    }
};

void add( A obj1, B obj2)
{

int m, n;

m = obj1.a + obj2.x;
n = obj1.b + obj2.y;

cout<< "\nAddition of First data member of both object :"<<m;
cout<< "\n Addition of Second data member of both object :"<<n;

}

void main()
{
    A ob1(2,3);
    B ob2(10,10);

    add(ob1,ob2);

}
```

OUTPUT

```
Addition of First data member of both objects : 12
Addition of Second data member of both objects : 13
```

3.7. FRIEND CLASSES

Just as we have the possibility to define a friend function, we can also define a class as friend of another one, granting that second class access to the protected and private members of the first one. We have given an example to understand the concept of friend class.

Example 3.19 **Friend class**

```
#include<iostream.h>
class A
{
    int a, b;
public:
    A(int m, int n)
    {
        a = m;
        b = n;
    }

friend class B;
};
```

```
class B
{
    int x, y;
public:
    B(int m, int n)
    {
        x = m;
        y = n;
    }

    void add()
    {
        A objA(2, 3);

        x = x + objA.a;
        y = y + objA.b;
    }
    void putdata()
    {
        cout<< "\n The values of x :" << x <<" and y : "<<y;
    }
};

void main()
{
        B obj(10,10);

        obj.add();

        obj.putdata();
}
```

OUTPUT

```
The value of x : 12 and y : 13
```

Related Questions In Exams

Q.1. What do you understand by a static member function of a class? Discuss their characteristics. Give an example where you can justify the use of static member functions. *[May-2007]*

Ans. Refer section 3.4.3.

Q.2. What is friend function? Discuss its characteristics. *[May-2007]*

Ans. Refer Section 3.6.

Q.3. Explain, what is a friend class? *[Dec-2007] [May-2008]*

Write a program to demonstrate the concept of Friend class. *[Dec-2008]*

Ans. Refer Section 3.7.

Q.4. Explain the difference between inline function and friend function with the help of example. *[May-2008]*

Ans. Refer Section 3.5 and 3.6.

Q.5. Differentiate between the Public, private & protected access type? *[Dec-2007]*

Ans. Refer Section 3.2.2.

3.8. DYNAMIC MEMORY ALLOCATION

Until now, in all our programs, we have only had as much memory available as we declared for our variables, having the size of all of them to be determined in the source code, before the execution of the program. But, what if we need a variable amount of memory that can only be determined during runtime? For example, in the case that we need some user input to determine the necessary amount of memory space.

The answer is *dynamic memory*, for which C++ integrates the operators new and delete.

3.8.1 Operators New and New[]

In order to request dynamic memory we use the operator new. new is followed by a data type specifier and-if a sequence of more than one element is required-the number of these within brackets []. It returns a pointer to the beginning of the new block of memory allocated. Its form is :

```
pointer = new type-specifier
pointer = new type-specifier [elements]
```

The first expression is used to allocate memory to contain one single element of type-specifier typc. Thc second one is used to assign a block (an array) of elements of type-specifier type, where elements is an integer value representing the number of these. For example :

```
int * bobby;
bobby – new int [5];
```

In this case, the system dynamically assigns space for five elements of type int and returns a pointer to the first element of the sequence, which is assigned to bobby. Therefore, now, bobby points to a valid block of memory with space for five elements of type int.

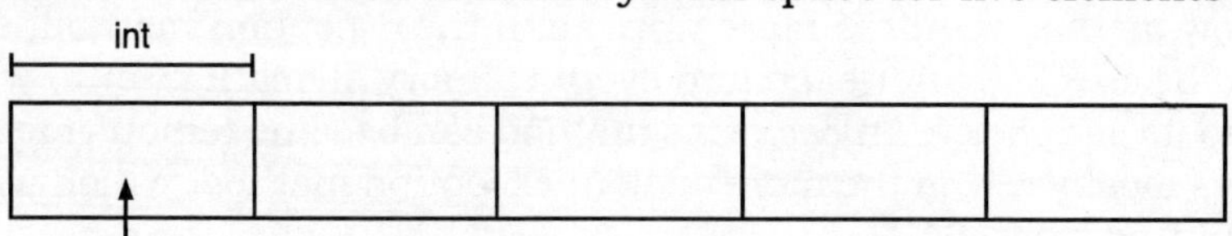

The first element pointed by bobby can be accessed either with the expression bobby[0] or the expression *bobby. Both are equivalent as has been explained in the section about pointers. The second element can be accessed either with bobby[1] or *(bobby+1) and so on.

You could be wondering the difference between declaring a normal array and assigning dynamic memory to a pointer, as we have just done. The most important difference is that the size of an array has to be a constant value, which limits its size to what we decide at the moment of designing the program, before its execution, whereas the dynamic memory allocation allows us to assign memory during the execution of the program (runtime) using any variable or constant value as its size.

The dynamic memory requested by our program is allocated by the system from the memory heap. However, computer memory is a limited resource, and it can be exhausted. Therefore, it is important to have some mechanism to check if our request to allocate memory was successful or not.

C++ provides two standard methods to check if the allocation was successful :

One is by handling exceptions. Using this method an exception of type bad_alloc is thrown when the allocation fails. Exceptions are a powerful C++ feature explained later in these tutorials. But for now you should know that if this exception is thrown and it is not handled by a specific handler, the program execution is terminated.

This exception method is the default method used by new, and is the one used in a declaration like :

```
bobby = new int [5]; // if it fails an exception is thrown
```

The other method is known as nothrow, and what happens when it is used is that when a memory allocation fails, instead of throwing a bad_alloc exception or terminating the program, the pointer returned by new is a null pointer, and the program continues its execution.

This method can be specified by using a special object called nothrow as parameter for new :

```
bobby = new (nothrow) int [5];
```

In this case, if the allocation of this block of memory failed, the failure could be detected by checking if bobby took a null pointer value :

```
int * bobby;
bobby = new (nothrow) int [5];
if (bobby = = 0)
{
    // error assigning memory. Take measures.
}
```

This nothrow method requires more work than the exception method, since the value returned has to be checked after each and every memory allocation, but I will use it in our examples due to its simplicity. Anyway this method can become tedious for larger projects, which nowadays mostly rely in the more eficient exception method, which will be explained in detail later in this tutorial.

3.8.2 Operator Delete and Delete[]

Since the necessity of dynamic memory is usually limited to specific moments within a program, once it is no longer needed it should be freed so that the memory becomes available again for other requests of dynamic memory. This is the purpose of the operator delete, whose format is :

```
delete pointer;
delete [] pointer;
```

The first expression should be used to delete memory allocated for a single element, and the second one for memory allocated for arrays of elements.

The value passed as argument to delete must be either a pointer to a memory block previously allocated with new, or a null pointer (in the case of a null pointer, delete produces no effect).

Example 3.20 **Dynamic memory program**

```
#include <iostream.h>
int main ()
{
    int i,n;
    int * p;
    cout << "How many numbers would you like to type? ";
    cin >> i;
    p= new (nothrow) int[i];
    if (p == 0)
        cout << "Error: memory could not be allocated";
    else
    {
        for (n=0; n<i; n++)
        {
            cout << "Enter number: ";
            cin >> p[n];
        }
        cout << "You have entered: ";
        for (n=0; n<i; n++)
            cout << p[n] << ", ";
        delete[] p;
}
return 0;
    }
```

OUTPUT

```
How many numbers would you like to type? 5
Enter number : 75
Enter number : 436
Enter number : 1067
Enter number : 8
Enter number : 32
You have entered: 75, 436, 1067, 8, 32,
```

Notice how the value within brackets in the new statement is a variable value entered by the user (i), not a constant value :

```
p= new (nothrow) int[i];
```

But the user could have entered a value for i so big that our system could not handle it. For example, when I tried to give a value of 1 billion to the "How many numbers" question, my system could not allocate that much memory for the program and I got the text message we prepared for this case (Error: memory could not be allocated). Remember that in the case that we tried to allocate the memory without specifying the nothrow parameter in the new expression, an exception would be thrown, which if it's not handled terminates the program.

It is a good practice to always check if a dynamic memory block was successfully allocated. Therefore, if you use the nothrow method, you should always check the value of the pointer returned. Otherwise, use the exception method, even if you do not handle the exception. This way, the program will terminate at that point without causing the unexpected results of continuing executing a code that assumes a block of memory to have been allocated when in fact it has not.

3.9 THIS POINTER

C++ provides a keyword "this", which represents an object that invokes a member function. Whenever any object called its member function this pointer is automatically set and contains the address of that object. The pointer acts as an implicit argument to all the member function. For example :

```
class xyz
    {
        float a;
        int b;
        ....
        ....
        ....
        };
```

The private member a and b can be used directly inside any member function of class xyz as follows :

```
a = 6.75;
b = 24;
```

We can also use the following satement inside the member funtion of class xyz :

```
this-> a = 6.75;
this-> b = 24;
```

We are generally not used this form because we have previouse shorthand form, one important application of "this" is to return the object, it points to as follows :

```
return *this;
```

Here we given an example to understand the concept of "this" pointer.

Example 3.21 This pointer

```
#include<iostream.h>
#include<string.h>
class student
{
    int rollno;
    int total;
    char name[10];
public:
    student (int rln, int tot, char *str)
    {
        rollno = rln;
        total = tot;
        strcpy ( name, s);
    }
student() { }
```

```
void putdata()
{
    cout<<"\nRoll No :"<<rollno;
    cout<<"\n Name :"<<name;
    cout<<"\n Total Marks :"<<total;
}
student & topper (student &stu)
{
    if (st.total > total)
       return st;
    else
       return *this;
}
};

void main()
{
    student S1(5, 450, "Amit");
    student S2(6, 425, "Rajkumar");
    student S3;
    student S4(7, 500, "Harsh");

    S3 = S1.topper(S2);
    cout<<"\n Topper in S1 and S2 is :"<<endl;
    S3.putdata();

    S3 = S1.topper(S4);
    cout<<"\n Topper in S1 and S4 is :"<<endl;
    S3.putdata();
}
```

OUTPUT

```
Topper in S1 and S2 is :
Roll No: 5
Name : Amit
Total Marks : 450

Topper in S1 and S4 is :
Roll No: 7
Name : Harsh
Total Marks : 500
```

Related Questions In Exams

Q.1. How the dynamic memory is allocated? *[Dec-2007]*

Ans. Refer Section 3.7.1.

Q.2. What is the use of new and delete operator? *[May-2007]*

Ans. Refer Section 3.7.1 and 3.7.2.

Q.3. What is this pointer? How can we use it? *[Dec-2008]*

Ans. Refer Section 3.8.

Q.4. What is the difference between new/delete and malloc/free? *[Dec-2009]*

Ans.

new/delete	*malloc/free*
1. These are operators.	1. These are functions.
2. It automatically computes the size to allocate and release memory.	2. It manually required size to allocate and release the memory.
3. It automatically returns the correct pointer type.	3. It needs to specify the returns pointer type.
4. New/delete can be overload.	4. Overloading is not possible.

3.10 CONTAINER CLASSES

In real life, we use containers all the time. Your breakfast food comes in a box, the pages in your book come inside a cover and binding, and you might store any number of items in containers in your garage. Without containers, it would be extremely inconvenient to work with many of these objects. Imagine trying to read a book that didn't have any sort of binding, or eat food that didn't come in a box without using a bowl. It would be a mess. The value the container provides is largely in it's ability to help organize and store items that are put inside it.

Similarly, a **container class** is a class designed to hold and organise multiple instances of another class. There are many different kinds of container classes, each of which has various advantages, disadvantages, and restrictions in their use. By far the most commonly used container in programming is the array, which you have already seen many examples of. Although C++ has built-in array functionality, programmers will often use an array container class instead because of the additional benefits it provides. Unlike built-in arrays, array container classes generally provide dynamically resizing (when elements are added or removed) and do bounds-checking. This not only makes array container classes more convenient than normal arrays, but safer too.

As you know class may have collection of different data types as data member, therefore it may have objects of other classes as data member. When any class contains an object of another class as a member, this class is called a container class. See example :

```
class X
{
    :
    :
};

class Y
{
    :
    :
public :
    X  obj;
};
```

In contrast, a class Y that has a member of another class X is said to have X or Y contain X. This relationship is called a membership or a "has a" relationship.

Definition : When any class contains one or more objects of another classes as member, this class is called a **container** class.

Container classes typically implement a fairly standardised minimal set of functionality. Most well-defined containers will include functions that :

- Create an empty container (*via.,* a constructor)
- Insert a new object into the container
- Remove an object from the container
- Report the number of objects currently in the container
- Empty the container of all objects
- Provide access to the stored objects
- Sort the elements (optional)

Sometimes certain container classes will omit some of this functionality. For example, arrays container classes often omit the insert and delete functions because they are slow and the class designer does not want to encourage their use.

Container classes generally come in two different varieties.

- **Value containers** are compositions that store copies of the objects that they are holding (and thus are responsible for creating and destroying those copies).
- **Reference containers** are aggregations that store pointers or references to other objects (and thus are not responsible for creation or destruction of those objects).

Unlike in real life, where containers can hold whatever you put in them, in C++, containers typically only hold one type of data. For example, if you have an array of integers, it will only hold integers. Unlike some other languages, C++ generally does not allow you to mix types inside a container. If you want one container class that holds integers and another that holds doubles, you will have to write two separate containers to do this (or use templates, which is an advanced C++ feature). Despite the restrictions on their use, containers are immensely useful, and they make programming easier, safer, and faster.

3.10.1 When to use Container Classes

Consider the example of Jet_plane class. You might want to consider deriving a class for Jet_plane class from a class called *engine.* However a Jet_plane is not an engine but has an engine. To be able to decide whether to use inheritance or membership you can ask whether a Jet_plane has more than one engine. If that is possibility, it is most likely that the relationship will be "has a" rather than "is a" relationship. Not all relationship will fall under clear-cut categories. Therefore using a derived class or a contain class is a matter to be considered at the design stage. Let see an example :

Example 3.22

```
#include<iostream.h>
#include<conio.h>
class engine
{
    int num;
public :
    engine(int a){num=a;}
    void show()
```

```
        {
            cout<< "\n It is a Engine num "<<num;
        }
    };
    class jet_plane
    {
        engine e1(1)
        engine e2(2);
    public:
        void display_engine()
        {
            e1.show();
            e2.show();
        }
    };
    int main()
    {
        clrscr();
        jet_plane P;
        P.display_engine();
        return 0
    }
```

OUTPUT

```
It is Engine num 1
It is Engine num 2
```

3.10.2 Types of Container Classes

There are two types of container classes on the basis of containing objects.

1. Homogeneous container
2. Heterogeneous container

Homogeneous Container : When a container class contains a group of objects of a single class is called homogeneous container classes.

Heterogeneous Container : When a container class contains a group of mixed objects of different-different classes is called heterogeneous container classes.

3.11 PROXY CLASSES

A proxy is a class that provides a modified interface to another class. Here is an example - suppose we have an array class that we we only want to be able to contain the binary digist 1 or 0. Here is a first try :

```
struct array1 {
    int mArray[10];
    int & operator[]( int i) {
        /// what to put here
    }
};
```

We want operator[] to complain if we say something like a[1] = 42, but that isn't possible because the operator only sees the index of into the array, not the value being stored.

We can solve this using a proxy :

```
#include <iostream>
using namespace std;;
struct aproxy {
    aproxy( int & r ) : mPtr( & r ) {}
    void operator = ( int n ) {
        if ( n > 1 ) {
            throw "not binary digit";
        }
        *mPtr = n;
    }
    int * mPtr;
};
struct array {
    int mArray[10];
    aproxy operator[]( int i) {
        return aproxy( mArray[i] );
    }
};
int main() {
    try {
        array a;
        a[0] = 1;          // ok
        a[0] = 42;         // throws exception
    }
    catch( const char * e ) {
        cout << e << endl;
    }
}
```

The proxy class now does our checking for a binary digit and we make the array's operator[] return an instance of the proxy which has limited access to the array's internals.

Definition : A proxy class is a stand-in for another class.

Anotther example , Let's suppose you have a class that has a method that takes 60 seconds to complete. That means everytime you call that method, your program waits. But let's also assume you rarely call that method. Let's further assume this method is named Load() and the class is MyClass

```
class MyClass
{
    public:
void Load();             //takes a long time
void AMethod();
etc..                    //the other methods.
};.
```

The proxy class would look like :

```
class MyClassProxy
{
MyClass* theObject;
public:
```

```
MyClassProxy(); : theObject(0) {}
MyClass* operator->();
MyClass& operator*();
};
```

So when you create a MyClassProxy object, the MyClass* inside is set to zero.

Now you use MyClassProxy objects instead of MyClass objects.

If someone needs the MyClass object, they use the operator-> overload of MyClassProxy. This function just returns the MyClass* if the MyClass object exists otherwise is creates it and calls Load().

```
MyClass* MyClassProxy::operator->()
{
    if (!this->theObject)
    {
        theObject = new MyClass;
        theObject->Load();
    }
    return theObject;
};
```

So not until you use the proxy object with the -> operator do you see the 60 seccond delay.

```
MyClassProxy p;        //no delay
p->AMethod();          //Here the MyClass object is created
                       //Loaded and the MyClass::AMethod called.
```

3.12. FUNCTION OVERLOADING

The term *overloading* means 'providing multiple definitions of'.

Overloading of functions involves defining distinct functions which share the same name, each of which has a unique signature. When an overloaded function is called, the compiler decides which version of the function to use based upon the type and/or number of arguments, calling the function that has the closest match. Function overloading is appropriate for :

- Defining functions which essentially do the same thing, but operate on different data types.
- Providing alternate interfaces to the same function.

Function overloading is purely for a programming convenience.

Consider a function, GetTime, which returns in its parameter(s) the current time of the day, and suppose that we require two variants of this function: one which returns the time as seconds from midnight, and one which returns the time as hours, minutes, and seconds. Given that these two functions serve the same purpose, there is no reason for them to have different names.

C++ allows functions to be overloaded, that is, the same function to have more than one definition :

```
long GetTime (void);      // seconds from midnight
void GetTime (int &hours, int &minutes, int &seconds);
```

When GetTime is called, the compiler compares the number and type of arguments in the call against the definitions of GetTime and chooses the one that matches the call. For example :

```
int h, m, s;
    long t = GetTime();  // matches GetTime(void)
    GetTime(h, m, s);    // matches GetTime(int&, int&, int&);
```

To avoid ambiguity, each definition of an overloaded function must have a unique signature.

Member functions of a class may also be overloaded :

```
class Time
    {
    //... long GetTime (void);       // seconds from midnight
    void GetTime (int &hours, int &minutes, int &seconds);

    };
```

Function overloading is useful for obtaining flavors that are not possible using default arguments alone. Overloaded functions may also have default arguments :

```
void Error (int errCode, char
*errMsg = “”);
void Error (char *errMsg);
```

More than one function with same name and different arguments is function overloading.

Example 3.23 **Function overloading**

```
#include<iostream.h>
void myfunc(int a)
{
    cout << “a is “ << a << endl;
}
// overload myfunc
void myfunc(int a, int b)
{
    cout << “a is “ << a << endl;
    cout << “b is “ << b << endl;
}
// overload myfunc, again
void myfunc(int a, double b)
{
    cout << “a is “ << a << endl;
    cout << “b is “ << b << endl;
}
void main()
{
    myfunc(10);            // calls myfunc(int)
    myfunc(12, 24);        // calls myfunc(int, int)
    myfunc(99, 123.23);    // calls myfunc(int, double)
}
```

OUTPUT

```
a is 10
```

```
a is 12
b is 24

a is 99
b is 123.23
```

Example 3.24 **Overload area function**

```
#include<iostream.h>
# define PI 3.14
float area(float r)
{
    return PI*r*r;
}
float area(float l,float b)
{
    return (l*b)/2;
}
int area(int a)
{
    return a*a;
}
void main()
{
    float x = area(7.2f);
    float y = area(2.0f,8.4f);
    int z = area(2);
    cout<<"\nthe area of circle is="<<x;
    cout<<"\n the area of rectangle is ="<<y;
    cout<<"\nthe area of square is="<<z;
}
```

Output

```
the area of circle is = 162.7776
the area of rectangle is = 8.4
the area of square is = 4
```

Summary

Structure and class both are example of encapsulation. The data declared private in class or structure can't be access outside the class or structure. The difference between class and structure is that by default all member of class are private and in structure are public.

Constructors and destructors are the heart of classes and object from these two functions compiler create and destroy the memory for class objects.

Friend function and friend class allow private members to access outside the class.

The 'new' operator used to allocate dynamic memory to objects and 'delete' used to destroy them.

Function over loading provides new definitions for same name functions.

Solved Problems

Q. 3.1. In what ways does a class differ from a structure? in what ways are they similar?

Ans. A class in C++ is identical to structure in C++; the major difference lies in the default access category of its members. By default members are public in structure and private in class.

The members of structure in C++ are public by default, while the members of class are private by default.

Q. 3.2. Why we can't initialise data members within a structure or a class?

Ans. Because the declaration of structure or class reserves no memory, thus no memory exists that can store the value of a hypothetical initialisation statement. Therefore direct initialisation of the data members is not possible.

Q. 3.3. What is the difference between class declaration and definition?

Ans. A class declaration would be needed in the case where another calss needs to refer to the first class as a function argument, the compiler only needs to be told that the class name is valid. This can be done with a declaration.

see an example:

```
class A;          // class declartion of class A
class B           // class definition of class B
{
    ......
    ......
    ......
    void func( A obj); // object of class A is used before definition
};
class A
{
// members of class A;
    };
```

Q. 3.4. What is the scope resolution operator, and what its the role?

Ans. The scope resolution operator " : : " especially defined in C++. This may be use in two ways:

1. Binary Form: For resolving the class member function
2. Unary Form: For resolving the global variable.

1. Binary Form: When a member function is defined outside the class definition, it always has complete and unlimited access to the class's private member data. For example

```
class Demo : : display()
  {
        cout<<"\n How are you";
  }
```

2. Unary Form: You know that we can define more than one variable of same name in the different scopes(or blocks), the scope resolution operator is help to remove the ambiguty as well as provides the scope for variables wherever they needed. It also supports in the function overriding. we have given an example to understand the concept of unary form. For example

Example 3.1 Scope resolution

```
#include <iostream.h>
int num = 1;      // global variable num
```

```
class Demo
{
    int num;        // class variable num
public:
    Demo(int a) { num = a;}
    void display();
};

void Demo : : display()
{
    int num = 2;        // local variable num of display function
    cout<<num;          // it will print the value of local variable of
                           display function
    cout<< : : num;     // it will print the value of global variable
    cout<< Demo : : num;   // it will print the value of class variable
}
void main()
{
    Demo D(3);
    Demo.display();
}
```

Output of Example 3.1 Scope resolution

2

1

3

Q. 3.5. When would we define a member function inside or outside of the class definition?

Ans. If your function is very small and you need to make this inline, then you define your function in the class definition. because all defined function within class, are by default inline function. If you are using very much control statements and looping then you have to define your function outside the class.

Here's the basic idea:

- The public: part of the class body is where you describe the *observable semantics* of a class, its public member functions, its friend functions, and anything else exported by the class. Try not to provide any linkings of anything that can't be observed from the caller's code.
- The other parts of the class, including non-public: part of the class body, the definitions of your member and friend functions, etc. are pure implementation. Try not to describe any observable semantics that were not already described in the class's public: part.

Back to inlineness: the decision of whether a function is or is not inline is an implementation detail that does not change the observable semantics (the "meaning") of a call. Therefore the inline keyword should go next to the function's definition, not within the class's public: part.

Q. 3.6. How can inline functions help with the trade-off of safety *vs.* speed?

Ans. In straight C, you can achieve "encapsulated structs" by putting a void* in a struct, in which case the void* points to the real data that is unknown to users of the struct. Therefore users of the struct don't know how to interpret the stuff pointed to by the void*, but the access functions cast the void* to the appropriate hidden type. This gives a form of encapsulation.

Unfortunately it forfeits type safety, and also imposes a function call to access even trivial fields of the struct (if you allowed direct access to the struct's fields, anyone and everyone would be able to get direct access since they would of necessity know how to interpret the stuff pointed to by the void*; this would make it difficult to change the underlying data structure).

Function call overhead is small, but can add up. C++ classes allow function calls to be expanded inline. This lets you have the safety of encapsulation alongwith the speed of direct access. Furthermore the parameter types of these inline functions are checked by the compiler, an improvement over C's #define macros.

Q. 3.7. Why inline functions are mostly used instead of plain old #define macros?

Ans. Because #define macros are not good for use. Unlike #define macros, inline functions avoid infamous macro errors since inline functions always evaluate every argument exactly once. In other words, invoking an inline function is semantically just like invoking a regular function, only faster:

```
// A macro that returns the absolute value of i
#define unsafe(i)  ((i) >= 0 ? (i) : -(i) )
// An inline function that returns the absolute value of i
inline int safe(int i)
{
      return i >= 0 ? i : -i;
}
int f();
void userCode(int x)
{
      int ans;
      ans = unsafe(x++);    // Error! x is incremented twice
      ans = unsafe(f());    // Danger! f() is called twice
      ans = safe(x++);      // Correct! x is incremented once
      ans = safe(f());      // Correct! f() is called once
}
```

Also unlike macros, argument types are checked, and necessary conversions are performed correctly.

Q. 3.8. With inline member functions that are defined outside the class, is it best to put the inline keyword next to the declaration within the class body, next to the definition outside the class body, or both?

Ans. Best practice: only in the definition outside the class body.

```
class Foo {
public:
void method();   → best practice: don't put the inline keyword here
...
};
inline void Foo::method()  ← best practice: put the inline
keyword here
{ ... }
```

Q. 3.9. When writing a class constructor, how do you decide what to put in the initialisation and what to put in the body of the constructor?

Ans. A simple rule of thumb is that do as much as possible in the initialisation phase—means is, initialize all member variables there. Some things, like computations and print statements, must be in the body of the constructor.

Q. 3.10. Is there any difference between List x; and List x();?

Ans. A *big* difference!

Suppose that List is the name of some class. Then function f() declares a local List object called x:

```
void f()
{
List x;      // Local object named x (of class List)
...
}
```

But function g() declares a function called x() that returns a List:

```
void g()
{
List x();    // Function named x (that returns a List object)
...
}
```

Q. 3.11. Is the default constructor for Fred class always Fred::Fred()?

Ans. No. A "default constructor" is a constructor that *can be called* with no arguments. One example of this is a constructor that takes no parameters:

```
class Fred {
public:
Fred();    // Default constructor: can be called with no args
...
};
```

Another example of a "default constructor" is one that can take arguments, provided they are given default values:

```
class Fred {
public:
Fred(int i=3, int j=5);
                // Default constructor: can be called with args
...
};
```

Q. 3.12. Can I overload the destructor for my class?

Ans. No. You can have only one destructor. (For example, a class Fred. It's always called Fred::~Fred()). It never takes any parameters, and it never returns anything.

You can't pass parameters to the destructor anyway, since you never explicitly call a destructor (well, *almost* never).

Q. 3.13. Should I explicitly call a destructor on a local variable?

Ans. No! The destructor will get called *again* at the close } of the block in which the local variable was created. This is a guarantee of the language; it happens automatically; there's no way to stop it from happening. But you can get *really* bad results from calling a destructor on the same object a second time!.

Q. 3.14. Can an overloaded function have a default parameter?

Ans. Yes. There is no reason not to combine these powerful features. One or more of the overloaded functions can have their own default values, following the normal rules for default variables in any function.

Q. 3.15. Why are some member functions defined within the class declaration and others are not?

Ans. Defining the implementation of a member function within the declaration makes it inline. Generally, this is done only if the function is extremely simple. Note that you can also make a member function inline by using the keyword inline, even if the function is declared outside the class declaration.

Q. 3.16. Do friends violate encapsulation?

Ans. No! If they're used properly, they *enhance* encapsulation.

You often need to split a class in half when the two halves will have different numbers of instances or different lifetimes. In these cases, the two halves usually need direct access to each other (the two halves *used* to be in the same class, so you haven't *increased* the amount of code that needs direct access to a data structure; you've simply *reshuffled* the code into two classes instead of one). The safest way to implement this is to make the two halves friends of each other.

If you use friends like just described, you'll keep private things private. People who don't understand this often make naive efforts to avoid using friendship in situations like the above, and often they actually destroy encapsulation. They either use public data, or they make the data accessible between the halpes *via.,* public get() and set() member functions. Having a public get() and set() member function for a private data is OK only when the private data "makes sense" from outside the class (from a user's perspective). In many cases, these get()/set() member functions are almost as bad as public data: they hide (only) the *name* of the private data, but they don't hide the existence of the private data.

Similarly, if you use friend functions as a syntactic variant of a class's public access functions, they don't violate encapsulation any more than a member function violates encapsulation. In other words, a class's friends don't violate the encapsulation barrier: alongwith the class's member functions, they *are* the encapsulation barrier.

(Many people think of a friend function as something outside the class. Instead, try thinking of a friend function as part of the class's public interface. A friend function in the class declaration doesn't violate encapsulation any more than a public member function violates encapsulation: both have exactly the same authority with respect to accessing the class's non-public parts.)

Q. 3.17. Write a program to accept a date of birth and calculate the age of a person using classes.

Ans.
```
#include<iostream.h>
class DATE
{
     int dd, mm, yyyy;
public:
     void get_date()
     {
          cout<<"\n Enter Day :";
          cin>>dd;
          cout<<"\n Enter Month :";
          cin>>mm;
          cout<<"\n Enter Year :";
          cin>>yyyy;
     }
     void put_date()
     {
          cout<<dd<<"/"<<mm<<"/"<<yyyy;
     }
};
class Person
```

```
{
      char name[10];
      DATE DOB;
public:
      void getdata()
      {
            cout<<"\n Enter Your Name :";
            cin>>name;
            cout<<"\n Enter Your DoB :";
            DOB.get_date();
      }
      void putdata()
      {
            cout<<"\n Name :"<<name;
            cout<<"\n DoB :"<<DOB.put_date();
      }
      void find_age();
};
void Person :: find_age()
{
      DATE D;
      cout<<"\n Enter the Current Date:";
      D.get_date();
      if( D.yyyy < DOB.yyyy)
      {
            cout<<"\n Invalid Dates ";
            exit(1);
      }
      if( D.dd < DOB.dd)
      {
            D.dd += 30;
            D.mm--;
      }
 if( D.mm < DOB.mm)
      {
            D.mm += 12;
            D.yyyy--;
      }
      int d = D.dd - DOB.dd;
      int m = D.mm - DOB.mm;
      int y = D.yyyy - DOB.yyyy;
      cout<<"\n Your Age is ";
      cout<< y <<" years, "<< m <<" months, "<< d <<" and days.";
 }

 void main()
 {
      Person P;
```

```
        P.getdata();
        P.putdata();
        P.find_age();
    }
```

OUTPUT

```
        Enter Your Name :
        Amit
        Enter Your DoB :
        Enter Day :
        12
        Enter Month :
        3
        Enter Month :
        1978
        Name : Amit
        DoB : 12/3/1978
        Enter the Current Date :
        Enter Day :
        26
        Enter Month :
        5
        Enter Month :
        2005

    Your Age is : 27 years ,2 months, and 14 days
```

Q. 3.18. Write a program to perform following operations on STRING class:

(*a*) Input a string and find the length.

(*b*) Input a string and reverse .

(*c*) Input two strings and concatenate.

(*d*) Input two string and compare its equality

(*e*) Input a string and check it palindrome

Ans.

```
#include<iostream.h>
#include<string.h>
class STRING
{
    char str[10];
public:
    STRING(char x[])
    {
        strcpy(str,x);
    }
    void display()
    {
    cout<<str;
    }
    int str_len();
    STRING str_rev();
```

```
    voidstr_concat(STRING);
    int str_cmp();
    int str_palindrome();
};

int STRING :: str_len()
{
    for(int a=0; str[a]!='\0'; a++);

    return a;
}
STRING STRING :: str_rev()
{
    STRING X;
    int len=str_len();
    len--;
    for(int a=0;str[a] !='\0'; a++)
    X.str[a] = str[len--];

    return X;
}
void STRING :: str_concat(STRING X)
{
    char new_str[20];
    int a = 0;
    for(int b=0; str[b]!='\0'; b++)
    new_str[a++] = str[b];

    for(b=0; X.str[b]!='\0'; b++)
    new_str[a++] = X.str[b];

    new_str[a]='\0';
    cout<<"\n new string : "<<new_str;
}
int STRING :: str_cmp(STRING X)
{
    if(str_len()==X.str_len())
    {
    for(int a=0; str[a] != '\0'; a++)
    {
    if(str[a] != X.str[a])
            return 0;
    }
    return 1;
    }
    else
    return 0;
}
```

```
int STRING :: str_palindrome()
{
    STRING X;
    X = str_rev();
    if(str_cmp(X))
        cout<<"\n String is Palindrome";
    else
        cout<<"\n String is not Palindrome";
}
void main()
{
    STRING S("Computer");
    cout<<"\n Length of string is : "<<S.str_len();

    cout<<"\n Reverse of string is : ";

    STRING R;
    R = S.str_rev();
    R.display();
    S.str_concat(R);

    STRING P("Compute");
    if(S.str_cmp(P))
        cout<<"\n Both Strings are Same.";
else
    cout<<"\n Both strings are Not Same";

S.str_palindrome();
}
```

OUTPUT

```
    Length of string is : 7
    Reverse of string is : retupmoC
    new string : ComputerretupmoC
    Both strings are Not Same
String is not Palindrome
```

Q. 3.19. Create a class Time and write a program to add two time.

Ans.
```
#include <iostream.h>
class TIME
{
    int minutes, hours;
public:
    void get_time(int h, int m)
    {
        hours = h;
        minutes = m;
    }
```

```
    void put_time()
    {
        cout<<hours<<" : "<<minutes;
    }
    void add(TIME, TIME);
    };
    void TIME :: add(TIME t1, TIME t2)
    {
        minutes = t1.minutes + t2.minutes;
        hours = minutes / 60;
        minutes = minutes % 60;
        hours = hours + t1.hours + t2.hours;
    }

    void main()
    {
        TIME T1, T2, T3;

        T1.get_time(3,45);
        T2.get_time(6,34);
        T3.add(T1, T2);

        cout<<"\n Time 1 ="<<T1.put_time();
        cout<<"\n Time 2 ="<<T2.put_time();
        cout<<"\n Addition ="<<T3.put_time();
    }
```

OUTPUT

```
        Time 1 = 3:45
        Time 2 = 6:34
        Addition = 10:19
```

Q. 3.20. Create a class Complex for complex numbers and write a program to add two complex numbers using friend functions.

Ans.

```
#include<iostream.h>
    class Complex
    {
        float a,b;
    public:
        Complex(){ }
        Complex(float x)
        {
            a = b = x;
        }
        Complex(float real, float imag)
        {
            a = real;
            b = imag;
        }
        friend Complex add(Complex c1, Complex c2);
```

```
        friend void show(Complex);
    };
    Complex add(Complex c1, Complex c2)
    {
        Complex c3;
        c3.a = c1.a + c2.a;
        c3.b = c1.b + c2.b;
        return c3;
    }

    void show( Complex c)
    {
        cout<< c.a <<" + j"<<c.b<<endl;
    }

    void main()
    {
        Complex A(2.7, 3.5);
        Complex B(2.6);
        Complex C;

        cout<<"\n First Complex number :"<<show(A);
        cout<<"\n Second Complex number :"<<show(B);
        cout<<"\n Addition of both Complex numbers :";
        C = add(A,B);
        show(C);
    }
```

OUTPUT

```
        First Complex number : 2.7 + j3.5
        Second Complex number : 2.6 + j2.6
        Addition of both Complex numbers : 5.3 + j6.1
```

Q. 3.21. Create a class Matrix and write a program to construct a matrix where row and column input by user.

Ans.

```
#include<iostream.h>
    class Matrix
    {
        int **p;
        int row, col;
    public:
        Matrix (int, int);
        void get_data(int a, int b, int data)
        {
            p[a][b] = data;
        }

        int & put_data(int a, int b)
        {
            return p[a][b];
        }
```

```
};
Matrix :: Matrix(int a,int b)
{
     row = a;
     col = b;

     p = new int *[row];
     for( int i = 0; i<row ; i++)
          p[i] = new int[col];
}
void main()
{
     int m, n;
     cout<<"\n Enter the size of row :";
     cin>>m;
     cout<<"\n Enter the size of column :";
     cin>>n;

     Matrix M(m,n);
     cout<<"\n Enter the elements of Matrix:";
     int val;
     for(int a = 0; a < m; a++)
          for( int b = 0; b < n; b++)
          {
                cin>> val;
                M.get_data(a, b, val);
          }
cout<<"\nMatrix is :"<<endl;

for( a = 0; a < m; a++)
     {    for( int b = 0; b < n; b++)
          {
                cout<<M.put_data (a, b) <<"\t";
          }
     cout<<endl;
     }
}
```

OUTPUT

```
Enter the size of row :
3
Enter the size of column:
4
Enter the elements of Matrix:
11
12
13
14
15
```

```
16
17
18
19
20
21
22
Matrix is :
11  12   13   14
15  16   17   18
19  20   21   22
```

Exercise

Q. 3.1. What's difference between structure & class?

Q. 3.2. What is the role of constructor? Compare with destructor.

Q. 3.3. What are the types of constructors?

Q. 3.4. What is the role of copy constructors?

Q. 3.5. How static members of used with classes?

Q. 3.6. What is friend function? Give an example?

Q. 3.7. What is friend class? Give an example?

Q. 3.8. What is the role of new add delete operator?

Q. 3.9. What is the role of this pointer?

Q. 3.10. How container classes are useful for programme?

Q. 3.11. What is function overloading? Give an example?

❑❑❑

4

OPERATOR OVERLOADING

CHAPTER OUTLINE

- **Fundamentals**
- **Restrictions**
- **Overloading Unary Operator**
- **Overloading Binary Operator**
- **Overloading with Friend Function**
- **Summary**

4.1 FUNDAMENTALS

Operators are similar to functions in that they take operands (arguments) and return a value. Most of the built-in C++ operators are already overloaded.

For example, the + operator can be used to add two integers, two reals, or two addresses. Therefore, it has multiple definitions. The built-in definitions of the operators are restricted to built-in types.

For example :

```
int a, b,c;
float x, y, z;

c= a + b;        // '+' operator used for integers

z = x + y;       // '+' operator used for floats
```

Additional definitions can be provided by the programmer, so that they can also operate on user-defined types. Each additional definition is implemented by a function.

Here we given two programs to understand the method of using the operator overloading. First program is a simple program which adds two object's values and display these values. And second program adds two objects using "+" operator instead of add function.

Example 4.1 **Addition of two objects of same class**

```
#include<iostream.h>
class Demo
{
    int x, y;
public:
    Demo()
    {
        x = 0;
        y = 0;
    }
    Demo(int a, int b);
    {
        x = a;
        y = b;
    }
    Demo add(Demo);
    void putdata();
};

Demo Demo :: add(Demo D)
{
    Demo D1;

    D1.x = x + D.x;
    D1.y = y + D.y;

    return D1;
}

void Demo :: putdata()
{
    cout<<"\nThe value of x :"<< x << " and y :"<< y;
}

void main()
{
    Demo A;
    Demo B(3,2);
    Demo C(10,20);

    A = B.add(C) ;

    A.putdata();
}
```

OUTPUT

```
The value of x : 13 and y : 22
```

An operator may have multiple tasks using operator overloading. Such as : subtraction, multiplication and division can be done by + operator.

Example 4.2 **Addition of two objects of same class using operator overloading**

```
#include<iostream.h>
class Demo
{
    int x, y;
public:
    Demo()
    {
        x = 0;
        y = 0;
    }
    Demo(int a, int b);
    {
        x = a;
        y = b;
    }
    Demo add(Demo);
    void putdata();
};

// "+" operator is overloaded using the function definition

Demo Demo :: operator +(Demo D)
{
    Demo D1;

    D1.x = x + D.x;
    D1.y = y + D.y;

    return D1;
}

void Demo :: putdata()
    {
        cout<<"\nThe value of x :"<< x << " and y :"<< y;
    }

void main()
{
    Demo A;
    Demo B(3,2);
    Demo C(10,20);

    A = B + C;       // "+" is overloaded C is used as an argument

    A.putdata();
}
```

OUTPUT

```
The value of x : 13 and y : 22
```

4.2 RESTRICTIONS

Operator overloading is a best feature of the C++, it provides better arithmetic operation on the objects. But there is not possible to overload all the available operators in C++. The following list of operators which can be overloaded.

```
+      -    *    /    =    <    >    +=   -=   *=   /=   <<   >>
<<=  >>=  ==   !=   <=   >=   ++   --   %    &    ^    !    |
~      &=   ^=   |=   &&   ||   %=   []   ()   ->*  ->   new
delete  new[]    delete[]
```

As a C++ programmer or developer of C++ software you must know that there are some restriction on the operators to be overload.

Here we given some restriction of operator overloading :

- Precedence of an operator cannot be changed
- Associativity of an operator cannot be changed
- Arity (number of operands) cannot be changed
- Unary operators remain unary, and binary operators remain binary
- Operators &, *, + and-each have unary and binary versions
- Unary and binary versions can be overloaded separately
- No new operators can be created
  - Use only existing operators
- No overloading operators for built-in types
  - Cannot change how two integers are added
  - Produces a syntax error

There are some operators that **cannot** be overloaded.

```
sizeof     size of operator
.          membership operator
.*         Pointer-to-member operator
::         scope resolution operator
?:         Condition operator
```

There are some operators where a friend function **cannot** be used.

```
=          Assignment operator
()         function call operator]
[]         subscripting operator
->         class member access operator
```

4.3 OVERLOADING UNARY OPERATOR

Unary operators are the ones that operate on one operand, one such operator is the unary minus (–) operator which is used to change the sign of the operand it acts upon. This operator works well with basic data types such as int, float, etc.. In this section we will see how to overload this operator so that it can work the same way for user defined data types as it does for basic data types, *i.e.* change the sign of the operand.

The unary minus operator function does not take any arguments but it changes the sign of the data members of the object that calls this function. Since this function is a member function of the same class, it can directly access the members of the object that calls it.

Some Operator can not be overload, these are : . :: . ?: size of*

Example : -S

In the above statement, object S invokes the member function unary minus. In our example, the function will change the sign of all the data members of the object S.

A statement such as S1 =–S2 will yield a compiler error since our unary minus member function is not designed to return any value; however, with a little modification this function will be able to return values.

It is possible to overload a unary minus operator using a friend function as follows :

```
friend void operator-(space &s);        //declaration
void operator-(space &s)                //definition
{
    s.x =-s.x;
    s.y =-s.y;
    s.z =-s.z;
}
```

> **Note :**
> In the above example, we have received the operand by reference because we need to change the values of its data members, which would not be possible if we would receive the argument by value.

Example 4.3 **Overloading of the unary minus (–) operator.**

```
#include<iostream>
using namespace std;

class Minus
{
    private:
        int a, b, c ;
    public:
        Minus(int A, int B, int C)
        {
            a = A;
            b = B;
            c = C;
```

```
        }
        void display(void);
//********Declaration of the operator function**********
        void operator -  ( );
};

void Minus :: display(void)
{
    cout << "\t a = " << a << endl ;
    cout << "\t b = " << b << endl ;
    cout << "\t c = " << c << endl ;
}
//*********Definition of the operator function***********
inline void Minus :: operator-( )
{
    a =-a ;
    b =-b ;
    c =-c ;
}

//*************Main Function Definition***************
int main(void)
{
    Minus M(5, 10,-15) ;
    cout << "\n Before activating operator-( )\n" ;
    M.display( ) ;
    -M ;
    cout << "\n After activating operator-( )\n" ;
    M.display( ) ;
}
```

OUTPUT

```
Before activating operator-( )
    a = 5
    b = 10
    c =-15

After activating operator-( )
    a =-5
    b =-10
    c = 15
```

4.4 OVERLOADING BINARY OPERATOR

Binary operators can be overloaded in a similar manner as unary operators.

We should note the following features of an operator function for a binary operator :

1. It receives only one class type argument explicitly, in case of a member function. For a friend function, two class types are received as arguments.
2. It returns a class type.

Consider the following expression :

$$C = A + B;$$

The above expression contains three objects namely A, B and C of which A and B are the operands while C is the one to which the sum of A and B is assigned to. In the above expression, the operator function is supposed to add two user defined data types and return a user defined data type.

Now consider the statement :

temp.x = x + rhs.x;

In the above expression x is the data member Iof the object that is invoking the operator member function, while rhs.x refers to the data member of the object passed as an argument to the function. Since the function operator + ( ) needs to return the result, we need to have a temp variable (of the same class type) that stores the value to be returned.

It should be noted here that in overloading binary operators the object to the left of the operator is used to invoke the operator function while the operand to the right of the operator is always passed as an argument to the function.

Sample Operator Function

```
user_defined operator+(user_defined rhs)
{
    user_defined temp;
    temp.x = x + rhs.x;
    temp.y = y + rhs.y;
    return temp;
}
```

We can economise on vertical space by using only the following statement to perform the addition and the return :

```
return user_defined ((x+B.x),(y+B.y));      // works if appropriate
                                              constructor is defined
```

At this point you must be wondering how we can eliminate a temporary variable and use this statement instead. Well, this statement causes the appropriate constructor to be invoked, and thus initialises an object with no name which stores the result of the addition. This nameless object is then returned by the function. Using *temporary objects* in this way can make code shorter, more efficient and easier to read.

Example 4.4 **Overloading of a binary operators (+,–)**

```
#include<iostream>
using namespace std;

class Complex
{
private:
```

```
        float real, imaginary ;
    public:
        Complex( )  // Default Constructor without arguments
        { real = 0.0; imaginary = 0.0; }
        Complex(float r, float i) // Constructor with arguments
        {
            real = r ;
            imaginary = i ;
        }
        void enterdata(void)
        {
            cout << "\nEnter the real part of the second Complex number:
            " ;
            cin >> real ;
            cout << "\nEnter the imaginary part of the second Complex
            number: " ;
            cin >> imaginary ;
        }
        void display(void)
        {
            if (imaginary >=0)
                cout << real << "+" << imaginary << "i" << endl ;
            else
                cout << real << imaginary << "i" << endl ;
        }
        // Operator Function Definition for overloading the binary
        operator+
        Complex operator +(Complex c)
        {
            Complex temp ;
            temp.real = real + c.real ;
            temp.imaginary = imaginary + c.imaginary ;
            return temp ;
        }
        // Operator Function Definition for overloading the binary
        operator-
        Complex operator-(Complex c)
        {
            Complex temp ;
            temp.real = real-c.real ;
            temp.imaginary = imaginary-c.imaginary ;
            return temp ;
        }
    };  // End of the class definition
```

```
int main(void)
{
    Complex c1(2.5, 4.5), c2 ;  // Object Definitions
    cout << "\nThe first Complex number i.e. c1 is: " ;
    c1.display( ) ;  // Displays the first Complex number
    c2.enterdata( ) ;
    cout << "\nThe second Complex number i.e. c2 is: " ;
    c2.display( ) ;  // Displays the second Complex number
    Complex c3 = c1 + c2 ;
    cout << "\nc1 + c2 is: " ;
    c3.display( );  // Displays the third Complex number
    Complex c4 = c1-c2 ;
    cout << "c2-c1 is: " ;
    c4.display( );   // Displays the fourth Complex number
}
```

OUTPUT

```
The first Complex number i.e. c1 is: 2.5+4.5i

Enter the real part of the second Complex number: 4

Enter the imaginary part of the second Complex number: 5

The second Complex number i.e., c2 is: 4+5i

c1 + c2 is: 6.5 + 9.5i
c2 - c1 is: -1.5- 0.5i
```

Related Questions In Exams (Most Important)

Q.1. What is operator overloading? *[May-2007]*

What do you understand by operator overloaded? *[May-2008]*

What is operator overloading? Explain the usefulness of operator overloading. *[Dec-2008]*

What is operator overloading? What are the rules that govern operator overloading? *[May-2009]*

What is operator overloading? *[Dec-2009]*

Ans. Refer Section 4.1.

Q.2. Differentiate between overloading of binary operator using friend function and without using friend function. *[May-2007]*

Ans. Using Friend Function : Using friend function a binary operator can compute private data for objects of more than one class.

Without Friend Function : Using member function without using friend a binary operator can compute private data for objects of only one than one class.

Let see an example where + operator is overloaded with friend function and * operator overloaded without friend function :

```
#include<iostream.h>
#include<conio.h>
class Y;            //pre declaration for compiler
class X
{
    int x1,x2;
public:
    X(int a,int b){x1=a;x2=b;}
    void show()
    {
        cout<<"data of X :"<<x1<<","<<x2;
    }
    void operator * (X B)
    {
        X C(0,0);
        C.x1=x1*B.x1;
        C.x2=x2*B.x2;
        C.show();
    }
    friend void operator +(X x,Y y);
};
class Y
{
    int y1,y2;
public:
    Y(int a,int b){y1=a;y2=b;}
    friend void operator +(X x,Y y);
};
void operator +(X x,Y y)
{
    int a=x.x1+y.y1;
    int b=x.x2+y.y2;
    cout<<"\nAddition of X and Y data are: "<<a<<","<<b;
}
int main()
{
    clrscr();
    X X1(5,10);
    X X2(2,3);
```

```
    X1*X2;       //call for * operator function without friend

    Y Y1(10,15);
    X1+Y1;       //call for + operator friend function
    return 0;
}
```

OUTPUT

```
data of X : 10, 30
Addition of X and Y data are: 15, 25
```

Q.3. Write a program to overload "+" operator using friend function to concatenate two strings. *[May-2007]*

Write a program to overload a binary operator. *[Dec-2008]*

Ans.

```
#include<iostream.h>
#include<string.h>
const int size 80
class string
{
    char str[size];
public:
string()
{
    strcpy(str," ");
}

string( char s[])
{
    strcpy(str, s);
}
void display()
{
    cout<<str<<endl;
}
string operator+ (string s)
{
    if( strlen(str) + strlen(s) < size)
    {
        string temp;
        strcpy(temp.str, str);
        strcat(temp.str,s);
        return temp;
    }
else
        cout<<"\n String Overflow";
}
```

```
};

void main()
{
    string s1 = "Hi User ";
    string s2 = "Have a Nice Day";
    string s3;

s1.display();
s2.display();
s3 = s1 + s2;
s3.display();
}
```

OUTPUT

```
Hi User
Have a Nice Day
Hi User Have a Nice Day
```

Q.4. What are restrictions on operator overloading? *[Dec-2007]*

List the operators which cannot be overloaded. *[May-2008]*

Ans. Refer Section 4.2.

Q.5. How the operator can be overload? *[Dec-2007]*

Ans. Refer Example 4.1.

Q.6. Write a program to overload "~" (tiled) operator to transpose of matrix. *[May-2007]*

Ans.

```
#include<iostream.h>
#include<conio.h>
class MAT
{
    int a[3][3];
public:
    void getdata();
    void show();
    void operator ~();
};
void MAT::getdata()
{
    cout<<"Enter elements for 3x3 matrix:"<<endl;
    for(int i=0;i<3;i++)
    for(int j=0;j<3;j++)
        cin>>a[i][j];
}
void MAT::show()
{
```

```
    cout<<"Elements of matrix:"<<endl;
    for(int i=0;i<3;i++)
    {for(int j=0;j<3;j++)
        cout<<a[i][j]<<"\t";
    cout<<endl;
    }
}
void MAT::operator ~()
{
    int b[3][3];
    for(int i=0;i<3;i++)
        for(int j=0;j<3;j++)
            b[j][i]=a[i][j];
    for(i=0;i<3;i++)
        for(j=0;j<3;j++)
            a[i][j]=b[i][j];
}

int main()
{
    clrscr();
    MAT M1;
    M1.getdata();
    M1.show();
    ~M1;
    cout<<"After transpose"<<endl;
    M1.show();
    return 0;
}
```

Q.7. Write a program to overload a unary operator. *[May-2009]*

Ans. Refer Example 4.3.

Q.8. How are prefix and postfix versions of ++( ) differentiate? Write a suitable program to explain the same. *[Dec-2009]*

Ans.

```
#include<iostream.h>
#include<conio.h>
class TEST
{
    int a;
public:
    TEST(int x){a=x;}
    void show();
    void operator ++();
```

```
    void operator ++(int);
};
void TEST::show()
{
    cout<<a<<endl;
}

void TEST::operator ++()
{
    a=a+1;
}
void TEST::operator ++(int nil)
{
    a=a+1;
}
int main()
{
    clrscr();
    TEST T1(10), T2(20);
    ++T1;
    cout<<"After Prefix"<<endl;
    T1.show();

    T2++;
    cout<<"After postfix"<<endl;
    T2.show();
    return 0;
}
```

4.5. OVERLOADING WITH FRIEND FUNCTION

Some operator overloading involves the use of *friend* functions. A friend function is just like a normal C function, but it is permitted to access private members of the class within which it is declared. The fact that it is a normal C function means that it does not have access to a **this** pointer, and also that it can be called without having to name a class that it operates on.

Example 4.5 **Overloading with Friend Function**

```
#include <iostream.h>
class Num
{
        friend Num operator+(Num ob, int i);
        friend Num operator+(int i, Num ob);
        int count;
public:
```

```
        Num(int cc=0){
            count=cc;
    }
        Num& operator=(int i);
    void Show(){
        cout <<count<<endl;
    }
};

Num& Num::operator=(int i){
    count=i;
    return *this;
}

Num operator+(Num ob, int i) // This handles ob+int.
{
    Num temp;
    temp.count=ob.count+i;
    return temp;
}

Num operator+(int i, Num ob) // This handles int+ob.
{
    Num temp;
    temp.count=ob.count+i;
    return temp;
}

main(void)
{
    Num obj;
    obj=10;
    obj.Show();
    obj=10+obj;
    obj.Show();
    obj=obj+12;
    obj.Show();
    return 0;
}
```

OUTPUT

```
10
20
32
```

Summary

Operator overloading is method to achive polymorphism in C++. Using operator overloading an operator can do more than one operation on the operands and objects.

In this chapter we show this on both unary and Binary operators.

Solved Problems

Q. 4.1. Write a program to overload a "+" operator to add two complex numbers.

Ans.
```
#include<iosteram.h>
class Complex
{
    float a,b;
public:
    Complex(){ }
    Complex(float real, float imag)
    {
        a = real;
        b = imag;
    }
    Complex operator+(Complex c);
    void show();
};
Complex Complex :: operator+ (Complex c)
{
    Complex temp;
    temp.a = a + c.a;
    temp.b = b + c.b;
    return temp;
}

void Complex :: show()
{
    cout<< a <<" + j"<<b<<endl;
}
void main()
{
    Complex A, B, C;
    A = Complex(2.7, 3.5);
    B = Complex(1.6, 2.7);
    C = A + B;

    cout<<"\n First Complex number :"<<A.show();
    cout<<"\n Second Complex number :"<<B.show();
    cout<<"\n Addition of both Complex numbers :"<<C.show();
}
```

OUTPUT

```
First Complex number : 2.7 + j3.5
Second Complex number : 1.6 + j2.7
Addition of both Complex numbers : 4.3 + j6.2
```

Q. 4.2. Create a class STRING and add two strings using "+" operators.

Ans.

```
#include<iostream.h>
#include<string.h>
const int size 80
class string
{
    char str[size];
public:
string()
{
    strcpy(str," ");
}

string( char s[])
{
    strcpy(str, s);
}
void display()
{
    cout<<str<<endl;
}
string operator+ (string s)
{
    if( strlen(str) + strlen(s) < size)
    {
        string temp;
        strcpy(temp.str, str);
        strcat(temp.str,s);
        return temp;
    }
else
        cout<<"\n String Overflow";
}
};

void main()
{
    string s1 = "Hi User ";
    string s2 = "Have a Nice Day";
    string s3;

s1.display();
s2.display();
s3 = s1 + s2;
```

```
    s3.display();
}
```

OUTPUT

```
    Hi User
    Have a Nice Day
    Hi User Have a Nice Day
```

Q. 4.3. Create a class Distance to store the distances in feet and inches and overload '+' operator to add two and three distances.

Ans.

```
#include<iostream.h>
    class distance
    {
        int feet;
        float inches;
    public:
        distance()
        {
            feet = 0;
            inches = 0.0;
        }
        distance(int ft, float in)
        {
            feet = ft;
            inches = in;
        }
        void get_dist()
        {
            cout<<"\n Enter the feet :"<<endl;
            cin>>feet;
            cout<<"\n Enter the inches : "<<end;
            cin>>inches;
        }
        void display()
        {
            cout<<feet<<"\'-"<<inches<<'\"';
        }
        distance operator + (distance);
    };

    distance distance :: operator + (distance d2)
    {
        int f = feet + d2.feet;
        float i = inches + d2.inches;
        if( i >=12.0)
        {
            i = i - 12.0;
            f++;
            }
```

```
        return distance(f, i);
    }
    void main()
    {
        distance D1, D3, D4;
        D1.get_dist();
        distance D2(11, 6.25);
        D3 = D1 + D2;
        D4 = D1 + D2 + D3;

        cout<<"\n Distance 1 :"<<D1.display();
        cout<<"\n Distance 2 :"<<D2.display();
        cout<<"\n Distance 3 :"<<D3.display();
        cout<<"\n Distance 4 :"<<D4.display();
    }
```

OUTPUT

```
        Enter the feet :
        10
        Enter the inches :
        6.5
        Distance 1 : 10'-6.5"
        Distance 2 : 11'-6.25"
        Distance 3 : 22'-0.75"
        Distance 4 : 44'-1.5"
```

Q. 4.4. Create a class distance to store the distance in feet and inches and overload "<" or ">" operator to compare two distance.

Ans.

```
#include<iostream.h>
    class distance
    {
            int feet;
            float inches;
    public:
    distance ()
    {
            feet = 0;
            inches = 0.0;
    }
    distance (int ft, float in)
    {
            feet = ft;
            inches = in;
    }
    boolean operator > (distance);
    }
    boolean distance :: operator > (distance d2)
            {
```

```
        float bf1 = feet + inches/12;
        float bf2 = d2 feet + d2. inches /12;
        if (bf1 > bf2)
                return true;
        else
                return false;
    }
    void main ()
    {
        distance D1 (6, 2.5);
        distance D2(3, 3.5);
        if (D1 > D2)
                cout <<"distance D1 is greater",
        else
                cout << "distance D2 is greater";
    }
```

OUTPUT

```
    distance D1 is greater.
```

Q. 4.5. When should the keyword inline be used in a function prototype?

Ans. If the function is very small, no more than a line or two, and won't be called from many places in your program, it is a candidate for inlining.

Q. 4.6. What happens if we have the following two functions?

int Area (int width, int length = 1);

int Area (int size);

Will these overload? There are a different number of parameters, but the first one has a default value.

Ans. The declarations will compile, but if you invoke Area with one parameter you will receive a compile- time error: ambiguity between Area(int, int) and Area(int).

Q. 4.7. How big is a class object?

Ans. A class object's size in memory is determined by the sum of the sizes of its member variables. Class methods don't take up room as part of the memory set aside for the object. Some compilers align variables in memory in such a way that two-byte variables actually consume somewhat more than two bytes. Check your compiler manual to be sure, but at this point there is no reason to be concerned with these details.

Q. 4.8. If I declare a class Cat with a private member itsAge and then define two Cat objects, Frisky and Boots, can Boots access Frisky's itsAge member variable?

Ans. No. While private data is available to the member functions of a class, different instances of the class cannot access each other's data. In other words, Frisky's member functions can access Frisky's data, but not Boots'. In fact, Frisky is a completely independent cat from Boots, and that is just as it should be.

Q. 4.9. Why shouldn't I make all the member data public?

Ans. Making member data private enables the client of the class to use the data without worrying about how it is stored or computed. For example, if the Cat class has a method GetAge(), clients of the Cat class can ask for the cat's age without knowing or caring if the cat stores its age in a member variable, or computes its age on the fly.

Q. 4.10. If using a const function to change the class causes a compiler error, why shouldn't I just leave out the word const and be sure to avoid errors?

Ans. If your member function logically shouldn't change the class, using the keyword const is a good way to enlist the compiler in helping you find silly mistakes. For example, GetAge() might have no reason to change the Cat class, but your implementation has this line:

```
if (itsAge = 100)
cout << "Hey! You're 100 years old\n";
```

Declaring GetAge() to be const causes this code to be flagged as an error. You meant to check whether its Age is equal to 100, but instead you inadvertently assigned 100 to its Age. Because this assignment changes the class—and you said this method would not change the class—the compiler is able to find the error.

This kind of mistake can be hard to find just by scanning the code. The eye often sees only what it expects to see. More importantly, the program might appear to run correctly, but its Age has now been set to a bogus number. This will cause problems sooner or later.

Q. 4.11. Is there ever a reason to use a structure in a C++ program?

Ans. Many C++ programmers reserve the struct keyword for classes that have no functions. This is a throwback to the old C structures, which could not have functions. Frankly, we find it confusing and poor programming practice.

Exercise

Q. 4.1. What is operator overloading?

Q. 4.2. What operator are available for overloading and what are restricted?

Q. 4.3. Give an example of Unary operator overloading.

Q. 4.4. Give an example of Binary operator overloading.

❑❑❑

5

INHERITANCE

CHAPTER OUTLINE

5.1 INTRODUCTION

One of the most powerful features of classes in C++ is that they can be extended through inheritance. The feature of reusability, implements by the *inheritance* in C++ language.

Inheritance means taking an existing class and adding functionality by deriving a new class from it.

The class you start with is called the *base class*, and the new class you create is called the *derived class*.

When you derive a class from another class, the new class gets all the functionality of the base class plus whatever new features you add. You can add data members and functions to the new class, but you cannot remove anything from what the base class offers.

5.2 BASE AND DERIVED CLASSES

A derived class can be defined with specifying its relationship with the base class. The syntax and general form of derived class is :

```
class derived-class-name : derivation-mode base-class-name
{
    members of class
};
```

After the colon (:), there are two items, one is derivation-mode and second is base-class-name. The base-class-name indicates that the derived class is derived form this base class. The derivation-mode is optional and, if present, may be public, protected or private. The default derivation-mode is private. Derivation mode specifies whether the features of the base class are privately derived, protected or publicly derived. For example :

```
class Demo : private Base       // private derivation
    {
        members of class Demo
    };
class Demo : protected Base     // protected derivation
    {
        members of class Demo
    };
class Demo : public Base        // public derivation
    {
        members of class Demo
    };
class Demo : Base               // private derivation by default
    {
        members of class Demo
    };
```

5.3 MODE OF INHERITANCE

The default derivation-mode is private. Derivation mode specifies whether the features of the base class are privately derived, protected or publicly derived.

We are explaining the following types of derivation mode :

- Public
- Private
- Protected

5.3.1 Public Derivation Mode

If the derivation mode is public then the public and protected both members of base class will be inherit but private members will not inherit. Check the figure.

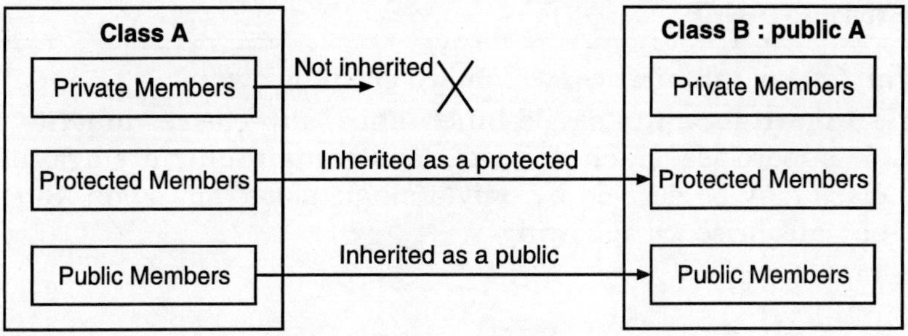

Fig. 5.1. Public derivation mode

5.3.2 Protected Derivation Mode

If the derivation mode is protected then the public and protected both members of base will be inherit as protected but private members will not inherit. Both the public and protected members will become protected for derived class. Check the figure.

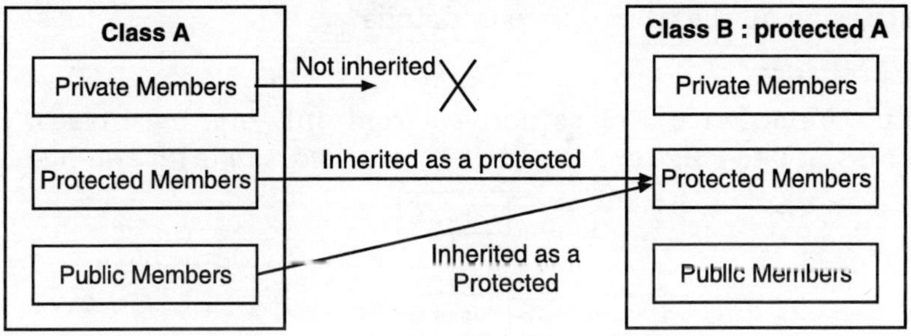

Fig. 5.2. Protected derivation mode

5.3.3 Private Derivation Mode

If the derivation mode is private then the public and protected both members of base will be inherit as private but private members will not inherit. Both the public and protected members will become private members for derived class. Check the figure.

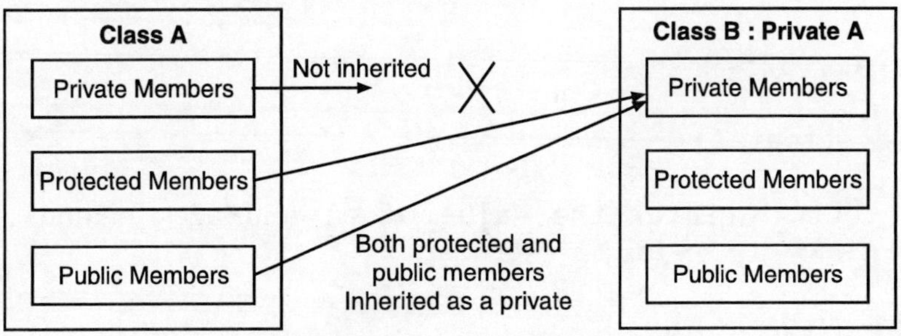

Fig. 5.3. Private derivation mode

The summarised rules for derived classes is shown in the following table :

Base Class	**Private Derived**	**Public Derived**	**Protected Derived**
Private Member	*Not inherited*	*Not inherited*	*Not inherited*
Public Member	*private*	*public*	*Protected*
Protected Member	*private*	*protected*	*Protected*

5.4 TYPES OF INHERITANCE

C++ supports all type of inheritance, which are given below :

1. **Single Inheritance :** In single Inheritance, sub-classes inherits the features of a single superclass. Each derived class contain only a single base class, and base class may be derived by only a single class and so on. Single inheritance may be categorise by following :
   (*a*) Simple
   (*b*) Multi-level
   (*c*) Hierarchical
2. **Multiple Inheritance :** Multiple inheritance allows a class to have more than one superclasses and to inherit features from all parent classes (or Base classes). Multiple inheritance may be categorise by following :
   (a) Multiple
   (*b*) Hybrid

We are discussing all the inheritance in details.

5.4.1 Simple Inheritance

In the simple inheritance a class derived from only one base class. If a class B is derived form class A, like a figure 11.1, then it is called *single inheritance.*

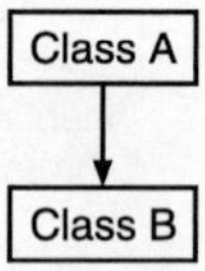

Fig. 5.4. Simple Inheritance

Here we have given some example to understand the concept of simple inheritance.

Example 5.1 **Simple inheritance**

```
#include<iostream.h>
class A
{
    int a1, a2;
public:
    void getdata()
    {
        cout<<"\nEnter the value of a1 and a2 :"<<endl;
        cin>> a1 >> a2;
    }
    void putdata()
    {
        cout<<"\nThe value of a1 is :"<< a1 << "and a2 :"<<a2;
    }
};

class B : public A        // class B is publicly derived by class
                          A,(single inheritance)
{
    int b1, b2;
```

```
public:
    void indata()
    {
        cout<<"\nEnter the value of b1 and b2 :" <<endl;
        cin>> b1 >> b2;
    }
    void outdata()
    {
        cout<<"\nThe value of b1 is :"<< b1 << "and b2 :"<<b2;
    }
};

void main()
{

    B Obj;

    Obj.getdata();          // base class member function

    Obj.indata();           // derived class member function

    Obj.putdata();
    Obj.outdata();
}
```

OUTPUT

```
Enter the value of a1 and a2 :
2
3

Enter the value of b1 and b2 :
4
5

The value of a1 is : 2 and a2 is : 3

The value of b1 is : 4 and b2 is : 5
```

Example 5.2 **Simple inheritance**

```
#include<iostream.h>

class First
{
protected:
    int a;          // a is protected variable will inherit in
                    derived class
public:
    void getdata()
    {
        cout<<"\nEnter the value of a :" <<endl;
```

```
        cin>> a;
    }
    void putdata()
    {
        cout<<"\nThe value of a is :"<< a;
    }
};

class Second : public First
{
    int b;
public:
    void indata()
    {
        cout<<"\nEnter the value of b :" <<endl;
        cin>> b;
    }
    void outdata()
    {
        cout<<"\nThe value of b is :"<< b;
    }

    void addition ()
    {
        int c = a+ b;
        cout<<"\n Addition of both variables: "<< c;
    }
};

void main()
{

    Second S;

    S.getdata();        // base class member function

    S.indata();         // derived class member function

    S.putdata();
    S.outdata();
    S.addition();
}
```

OUTPUT

```
Enter the value of a :
2

Enter the value of b :
4
```

```
The value of a is : 2

The value of b is : 4

Addition of both variables : 6
```

Example 5.3 **Simple inheritance**

```
#include<iostream.h>

class Student
{
    char name[10];
    int rollno;
public:
    void getdata()
    {
       cout<<"\nEnter the name :"<<endl;
       cin>> name;
       cout<< "\nEnter the Roll No. :"<<endl;
       cin>>rollno;
    }
    void putdata()
    {
       cout<<"\n Name :"<< name;
       cout<<"\n Roll No. :"<<rollno;
    }
};

class Marks : public Student
{
    int marks[4];
public:
    void getmarks();
    void putmarks();
};

void Marks :: getmarks()
{
    for(int i = 0; i <= 4 ; i++)
    {
       cout<<"\nEnter the Marks of Subject "<< i+1<< ": =" <<endl;
       cin>> marks[i];
    }
}
void Marks::putmarks()
{
    int total = 0;
    for(int i = 0; i <= 4 ; i++)
    {
```

```
        cout<<"\nMarks of Subject "<< i+1<< ": =" << marks[i];
        total = total + marks[i];
    }
    cout<<"\n Total Marks :"<<total;
}

void main()
{
    Marks M;

    M.getdata();
    M.getmarks();

    M.putdata();
    M.putmarks()
}
```

OUTPUT

```
Enter the Name :
Ravi

Enter the Roll No. :
1104

Enter the Marks of Subject 1
65

Enter the Marks of Subject 2
69

Enter the Marks of Subject 3
76

Enter the Marks of Subject 4
54

Name : Ravi

Roll No.: 1104
Marks of Subject 1
65

Marks of Subject 2
69

Marks of Subject 3
76
```

```
Marks of Subject 4
54

Total Marks : 264
```

5.4.2. Multi-level Inheritance

In the Multi-level inheritance a class derived from another derived class. If a class B is derived form class A and class C derived from class B, like a figure 5.5, then it is called *multilevel inheritance.* Because there are more than one levels. You can also derive class C, and then derive its derived and so on.

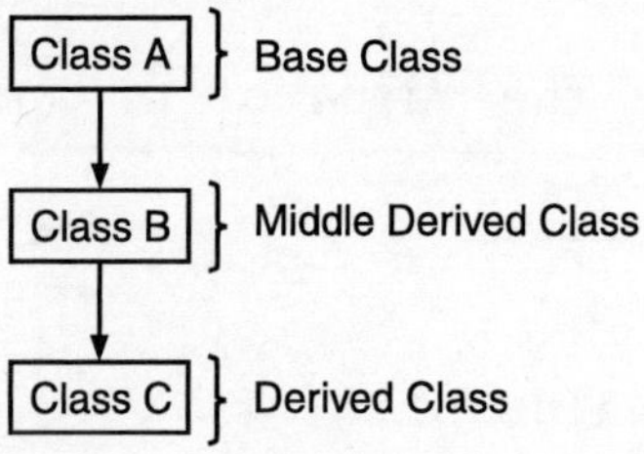

Fig. 5.5. Multi-level Inheritance

Here we have given some example to understand the concept of Multi-level inheritance.

Example 5.4 **Multi-level inheritance**

```
#include<iostream.h>

class A
{
    int a1, a2;
public:
    void getdata()
    {
        cout<<”\nEnter the value of a1 and a2 :”<<endl;
        cin>> a1 >> a2;
    }
    void putdata()
    {
        cout<<”\nThe value of a1 is :”<< a1 << “and a2 :”<<a2;
    }
};

class B : public A
{
    int b1, b2;
public:
    void indata()
    {
        cout<<”\nEnter the value of b1 and b2 :” <<endl;
        cin>> b1 >> b2;
    }
    void outdata()
```

```
    {
        cout<<"\nThe value of b1 is :"<< b1 << "and b2 :"<<b2;
    }
};

class C : public B
{
    int c1, c2;
public:
    void input()
    {
        cout<<"\nEnter the value of c1 and c2 :" <<endl;
        cin>> c1 >> c2;
    }
    void output()
    {
        cout<<"\nThe value of c1 is :"<< c1 << "and c2 :"<<c2;
    }
};

void main()
{

    C Obj;

    Obj.getdata();          // member function of class A

    Obj.indata();           // member function of class B

    Obj.input();            // member function of class C

    Obj.putdata();          // member function of class A
    Obj.outdata();          // member function of class B

    Obj.output();           // member function of class C

}
```

OUTPUT

```
Enter the value of a1 and a2 :
2
3

Enter the value of b1 and b2 :
4
5

Enter the value of c1 and c2 :
6
7
```

```
The value of a1 is : 2 and a2 is : 3

The value of b1 is : 4 and b2 is : 5

The value of c1 is : 6 and c2 is : 7
```

Example 5.5 **Multi-level inheritance**

```
#include<iostream.h>

class First
{
protected:
    int a;          // a is protected variable will inherit in
                    derived classes
public:
    void getdata()
    {
        cout<<"\nEnter the value of a :" <<endl;
        cin>> a;
    }
    void putdata()
    {
        cout<<"\nThe value of a is :"<< a;
    }
};

class Second : public First
{
protected:
    int b;          // b is protected variable will inherit in
                    derived classes
public:
    void indata()
    {
        cout<<"\nEnter the value of b :" <<endl;
        cin>> b;
    }
void outdata()
    {
        cout<<"\nThe value of b is :"<< b;
    }

};

class Third : public Second
{
    int c;
public:
```

```
    void input()
    {
        cout<<"\nEnter the value of c :" <<endl;
        cin>> c;
    }
void output()
    {
        cout<<"\nThe value of c is :"<< c;
    }

    void addition ()
    {
        int d = a + b + c;
        cout<<"\n Addition of variables: "<< d;
    }
};

void main()
{

    Third T ;

    T.getdata();      // member function of class First

    T.indata();       // member function of class Second

    T.input()         // member function of class Third

    T.putdata();
    T.outdata();
    T.output();

    T.addition();
}
```

OUTPUT

```
Enter the value of a :
2

Enter the value of b :
4

Enter the value of c :
6

The value of a is : 2

The value of b is : 4
```

```
The value of b is : 6

Addition of both variables : 12
```

Example 5.6 **Multi-level inheritance**

```
#include<iostream.h>
#include<string.h>

class Student
{
    char name[10];
    int rollno;
public:
    void getdata()
    {
        cout<<"\nEnter the name :"<<endl;
        cin>> name;
        cout<< "\nEnter the Roll No. :"<<endl;
        cin>>rollno;
    }
    void putdata()
    {
        cout<<"\n Name :"<< name;
        cout<<"\n Roll No. :"<<rollno;
    }
};

class Marks : public Student
{
    int marks[4];
protected:
    int total;
public:
    void getmarks();
    void putmarks();
};

void Marks :: getmarks()
    {
        for(int i = 0; i <= 4 ; i++)
        {
            cout<<"\nEnter the Marks of Subject "<< i+1<< ": ="
            <<endl; cin>> marks[i];
        }
    }
void Marks::putmarks()
    {
        total = 0;
        for(int i = 0; i <= 4 ; i++)
```

```
        {
            cout<<"\nMarks of Subject "<< i+1<< ": =" << marks[i];
            total = total + marks[i];
        }
        cout<<"\n Total Marks :"<<total;
}

class Result : public Marks
{
    int perc;
    boolean pass;           // it will store boolean value TRUE for
                            pass and FALSE for fail
    char division[10];
public:
    void check();
    void display();
};

void Result :: ckeck()
{
    pass = TRUE;
    for(int i = 0; i <= 4 ; i++)
    {
        if (marks[i]<30)
            pass = FALSE;
    }
        perc = (total * 100) / 400;
}

void Result :: display()
{
    if (pass = = TRUE)
    {
        if(perc >= 60)
            strcpy(division, "First");
        elseif(perc >= 45)
            strcpy(division, "Second");
        elseif(perc >= 30)
            strcpy(division, "Third");
        else
            strcpy(division, "Fail");
    }
    else
    {
        cout<<"\nStudent is Fail";
        break;
    }

    cout<<" \n Percentage :"<<perc<< "%";
```

```
        cout<<"\n Division :"<<division;
    }
    void main()
    {
        Result S;

        S.getdata();
        S.getmarks();

        S.putdata();
        S.putmarks()

        S.check();
        S.display();
    }
```

OUTPUT

```
    Enter the Name :
    Ravi

    Enter the Roll No. :
    1104

    Enter the Marks of Subject 1
    65

    Enter the Marks of Subject 2
    69

    Enter the Marks of Subject 3
    76

    Enter the Marks of Subject 4
    54

    Name : Ravi

    Roll No.: 1104

    Marks of Subject 1
    65

    Marks of Subject 2
    69

    Marks of Subject 3
    76

    Marks of Subject 4
    54
```

```
Total Marks : 264

Percentage : 66 %

Division : First
```

***Derived Class** inherits the features of **Base Class** like children acquire the features of their parents.*

5.4.3 Hierarchical Inheritance

In the Hierarchical inheritance more than one classes derived from only single base class. If a class B and class C is derived form class A, like a figure 5.6, then it is called hierarchical inheritance. Because there are more than one derived classes of a single base class thus both classes inherits the features of class A.

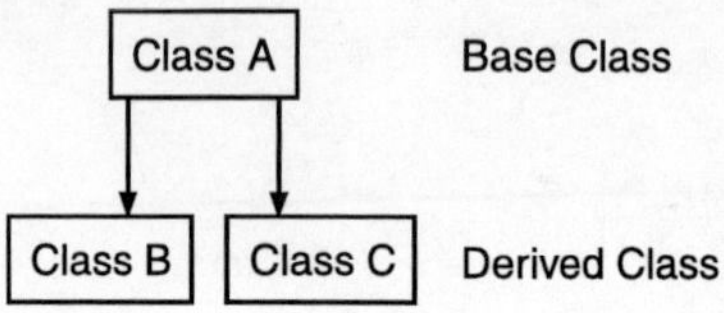

Fig. 5.6. Hierarchical Inheritance

Here we have given some example to understand the concept of Hierarchical inheritance.

Example 5.7 **Hierarchical inheritance**

```
#include<iostream.h>
class A
{
    int a1, a2;
public:
    void getdata()
    {
        cout<<"\nEnter the value of a1 and a2 :"<<endl;
        cin>> a1 >> a2;
    }
    void putdata()
    {
        cout<<"\nThe value of a1 is :"<< a1 << "and a2 :"<<a2;
    }
};

class B : public A
{
    int b1, b2;
public:
    void indata()
    {
        cout<<"\nEnter the value of b1 and b2 :" <<endl;
        cin>> b1 >> b2;
    }
    void outdata()
    {
```

```
        cout<<"\nThe value of b1 is :"<< b1 << "and b2 :"<<b2;
    }
  };

class C : public A
{
    int c1, c2;
public:
    void input()
    {
        cout<<"\nEnter the value of c1 and c2 :" <<endl;
        cin>> c1 >> c2;
    }
    void output()
    {
        cout<<"\nThe value of c1 is :"<< c1 << "and c2 :"<<c2;
    }
};

void main()
{

    B Ob1;
    C Ob2;

    Ob1.getdata();          // member function of class A
    Ob1.indata();           // member function of class B

    Ob2.getdata();          // member function of class A
    Ob2.input();            // member function of class C

    Ob1.putdata();          // member function of class A
    Ob1.outdata();          // member function of class B

    Ob2.putdata();          // member function of class A
    Obj.output();           // member function of class C
}
```

OUTPUT

```
Enter the value of a1 and a2 :
2
3

Enter the value of b1 and b2 :
4
5

Enter the value of a1 and a2 :
10
15
```

```
Enter the value of c1 and c2 :
6
7

The value of a1 is : 2 and a2 is : 3

The value of b1 is : 4 and b2 is : 5

The value of a1 is : 10 and a2 is : 15

The value of c1 is : 6 and c2 is : 7
```

Example 5.8 **Hierarchical inheritance**

```
#include<iostream.h>
class First
{
protected:
    int a;              // a is protected variable will inherit in
                        derived classes
public:
    void getdata()
    {
        cout<<"\nEnter the value of a :" <<endl;
        cin>> a;
    }
    void putdata()
    {
        cout<<"\nThe value of a is :"<< a;
    }
};

class Second : public First
{
int b;
public:
    void indata()
    {
        cout<<"\nEnter the value of b :" <<endl;
        cin>> b;
    }
    void outdata()
    {
        cout<<"\nThe value of b is :"<< b;
    }
    void addition ()
    {
        int d = a + b ;
        cout<<"\n Addition of variables: "<< d;
```

```
    }
};

class Third : public First
{
    int c;
public:
    void input()
    {
        cout<<"\nEnter the value of c :" <<endl;
        cin>> c;
    }
    void output()
    {
        cout<<"\nThe value of c is :"<< c;
    }

    void addition ()
    {
        int d = a + c;
        cout<<"\n Addition of variables: "<< d;
    }
};
void main()
{

    Second S;
    Third T ;

    S.getdata();      // member function of class First
    S.indata();       // member function of class Second

    T.getdata();      // member function of class First
    T.input()         // member function of class Third

    S.putdata();
    S.outdata();
    S.addition();

    T.putdata();
    T.output();
    T.addition();
}
```

OUTPUT

```
Enter the value of a :
2

Enter the value of b :
4
```

```
Enter the value of a :
10

Enter the value of c :
6

The value of a is : 2

The value of b is : 4

Addition of variables : 6

The value of a is : 10

The value of b is : 6

Addition of variables : 16
```

Example 5.9 **Hierarchical inheritance**

```
#include<iostream.h>
#include<string.h>

class Student
{
    char name[10];
    int rollno;
public:
    void getdata()
    {
        cout<<"\nEnter the name :"<<endl;
        cin>> name;
        cout<< "\nEnter the Roll No. :"<<endl;
        cin>>rollno;
    }
    void putdata()
    {
        cout<<"\n Name :"<< name;
        cout<<"\n Roll No. :"<<rollno;
    }
};

class Marks : public Student
{
    int marks[4];
    int total;
public:
    void getmarks();
    void putmarks();
```

```
};

void Marks :: getmarks()
    {
        for(int i = 0; i <= 4 ; i++)
        {
            cout<<"\nEnter the Marks of Subject "<< i+1<< ": ="
            <<endl; cin>> marks[i];
        }
    }
void Marks::putmarks()
{
    total = 0;
    for(int i = 0; i <= 4 ; i++)
    {
    cout<<"\nMarks of Subject "<< i+1<< ": =" << marks[i];
    total = total + marks[i];
    }
    cout<<"\n Total Marks :"<<total;
}

class Sports :public Student
{
    int score;
    char rank[20];
public:
    void getscore();
    void display();
};
void Sports :: getscore()
{
    cout<<"\nEnter the score of sports:"<<endl;
    cin>>score;
}
void Sports :: display()
{
        if(score >= 600)
            strcpy(rank, "First");
        elseif(perc >= 450)
            strcpy(rank, "Second");
        elseif(perc >= 300)
            strcpy(rank, "Third");
        else
            strcpy(rank, "Not Satsfatory");

        cout<<" \n Score :"<<score;
        cout<<"\n Rank :"<<rank;
}
void main()
```

```
{
    Marks M;
    Sports S;

    M.getdata();
    M.getmarks();

    S.getdata();
    S.getscore();

    M.putdata();
    M.putmarks()

    S.putdata();
    S.display();
}
```

OUTPUT

```
Enter the Name :
Ravi

Enter the Roll No. :
1104

Enter the Marks of Subject 1
65

Enter the Marks of Subject 2
69

Enter the Marks of Subject 3
76

Enter the Marks of Subject 4
54

Enter the Name :
Hari

Enter the Roll No. :
1136

Enter the score of Sports : 595

Name : Ravi

Roll No.: 1104

Marks of Subject 1
65
```

```
Marks of Subject 2
69

Marks of Subject 3
76

Marks of Subject 4
54

Total Marks : 264

Name : Hari

Roll No. : 1136

Score : 595

Rank : Second
```

These are used alternatively :
Parent ⇔ Base ⇔ Super class
Child ⇔ Derived ⇔ Sub class

5.4.4 Multiple Inheritance

In the Multiple inheritance a class derived from more than one base class. If a class C is derived form class A and class B, like a figure 5.7, then it is called multiple inheritance. Because there are more than one base classes thus class C inherits the features of class A and class B.

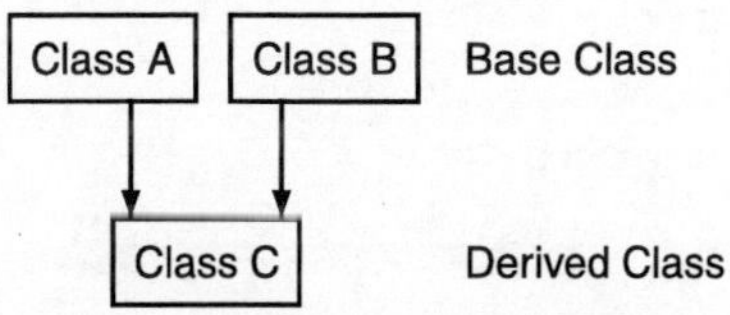

Fig. 5.7. Multiple Inheritance

Here we have given some example to understand the concept of Multiple inheritance.

Example 5.10 **Multiple inheritance**

```
#include<iostream.h>

class A
{
    int a1, a2;
public:
    void getdata()
    {
        cout<<"\nEnter the value of a1 and a2 :"<<endl;
        cin>> a1 >> a2;
    }
    void putdata()
    {
        cout<<"\nThe value of a1 is :"<< a1 << "and a2 :"<<a2;
    }
```

```
};

class B
{
    int b1, b2;
public:
    void indata()
    {
        cout<<"\nEnter the value of b1 and b2 :" <<endl;
        cin>> b1 >> b2;
    }
    void outdata()
    {
        cout<<"\nThe value of b1 is :"<< b1 << "and b2 :"<<b2;
    }
};

class C : public A, public B
{
    int c1, c2;
public:
    void input()
    {
        cout<<"\nEnter the value of c1 and c2 :" <<endl;
        cin>> c1 >> c2;
    }
    void output()
    {
        cout<<"\nThe value of c1 is :"<< c1 << "and c2 :"<<c2;
    }
};

void main()
{

    C Obj;

    Obj.getdata();        // member function of class A

    Obj.indata();         // member function of class B

    Obj.input();          // member function of class C

    Obj.putdata();        // member function of class A

    Obj.outdata();        // member function of class B

    Obj.output();         // member function of class C

}
```

OUTPUT

```
Enter the value of a1 and a2 :
2
3

Enter the value of b1 and b2 :
4
5

Enter the value of c1 and c2 :
6
7

The value of a1 is : 2 and a2 is : 3

The value of b1 is : 4 and b2 is : 5

The value of c1 is : 6 and c2 ´is : 7
```

Example 5.11 **Multiple inheritance**

```
#include<iostream.h>

class First
{
protected:
    int a;              // a is protected variable will inherit in
                        derived classes
public:
    void getdata()
    {
        cout<<"\nEnter the value of a :" <<endl;
        cin>> a;
    }
    void putdata()
    {
        cout<<"\nThe value of a is :"<< a;
    }
};

class Second
{
protected:
    int b;              // b is protected variable will inherit in
                        derived classes
public:
    void indata()
    {
        cout<<"\nEnter the value of b :" <<endl;
```

```
        cin>> b;
    }
    void outdata()
    {
        cout<<"\nThe value of b is :"<< b;
    }

};

class Third : public First, public Second
{
    int c;
public:
    void input()
    {
        cout<<"\nEnter the value of c :" <<endl;
        cin>> c;
    }
    void output()
    {
        cout<<"\nThe value of c is :"<< c;
    }

    void addition ()
    {
        int d = a + b + c;
        cout<<"\n Addition of variables: "<< d;
    }
};

void main()
{

    Third T ;

    T.getdata();          // member function of class First

    T.indata();           // member function of class Second

    T.input()             // member function of class Third

    T.putdata();
    T.outdata();
    T.output();

    T.addition();
}
```

OUTPUT

```
Enter the value of a :
```

```
2

Enter the value of b :
4

Enter the value of c :
6

The value of a is : 2

The value of b is : 4

The value of b is : 6

Addition of variables : 12
```

Example 5.12. **Multiple inheritance**

```
#include<iostream.h>
#include<string.h>

class Student
{
    char name[10];
    int rollno;
public:
    void getdata()
    {
        cout<<"\nEnter the name :"<<endl;
        cin>> name;
        cout<< "\nEnter the Roll No. :"<<endl;
        cin>>rollno;
    }
    void putdata()
    {
        cout<<"\n Name :"<< name;
        cout<<"\n Roll No. :"<<rollno;
    }
};

class Marks
{
    int marks[4];
protected:
    int total;
public:
    void getmarks();
    void putmarks();
};
```

```
void Marks :: getmarks()
{
    for(int i = 0; i <= 4 ; i++)
    {
        cout<<"\nEnter the Marks of Subject "<< i+1<< ": =" <<endl;
        cin>> marks[i];
    }
}
void Marks::putmarks()
{
    total = 0;
    for(int i = 0; i <= 4 ; i++)
{
        cout<<"\nMarks of Subject "<< i+1<< ": =" << marks[i];
        total = total + marks[i];
}
cout<<"\n Total Marks :"<<total;
}

class Result :public Student, public Marks
{
        int perc;
        boolean pass;     // it will store boolean value TRUE for
                          pass and FALSE for fail char division[10];
public:
        void check();
        void display();
};

void Result :: ckeck()
{
    pass = TRUE;
    for(int i = 0; i <= 4 ; i++)
    {
        if (marks[i]<30)
            pass = FALSE;
    }
    perc = (total * 100) / 400;
}

void Result :: display()
{
    if (pass = = TRUE)
    {
        if(perc >= 60)
            strcpy(division, "First");
        elseif(perc >= 45)
            strcpy(division, "Second");
        elseif(perc >= 30)
```

```
            strcpy(division, "Third");
        else
            strcpy(division, "Fail");
    }
    else
    {
        cout<<"\nStudent is Fail";
        strcpy(division, "Fail");
        break;
    }

    cout<<" \n Percentage :"<<perc<< "%";
    cout<<"\n Division :"<<division;
}
void main()
{
    Result S;

    S.getdata();
    S.getmarks();

    S.putdata();
    S.putmarks()

    S.check();
    S.display();
}
```

OUTPUT

```
Enter the Name :
Ravi

Enter the Roll No. :
1104

Enter the Marks of Subject 1
65

Enter the Marks of Subject 2
69

Enter the Marks of Subject 3
76

Enter the Marks of Subject 4
54

Name : Ravi

Roll No.: 1104
```

```
Marks of Subject 1
65

Marks of Subject 1
69

Marks of Subject 1
76

Marks of Subject 1
54

Total Marks : 264

Percentage : 66 %

Division : First
```

5.4.5 Hybrid Inheritance

A hybrid inheritance is a combination of two more inheritance. If two or more inheritance exists in an inheritance hierarchy, then this type of inheritance called *hybrid inheritance*.

Here we given some examples for hybrid inheritance.

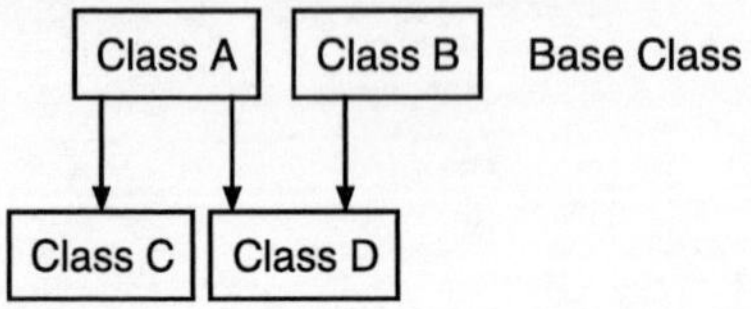

Fig. 5.8

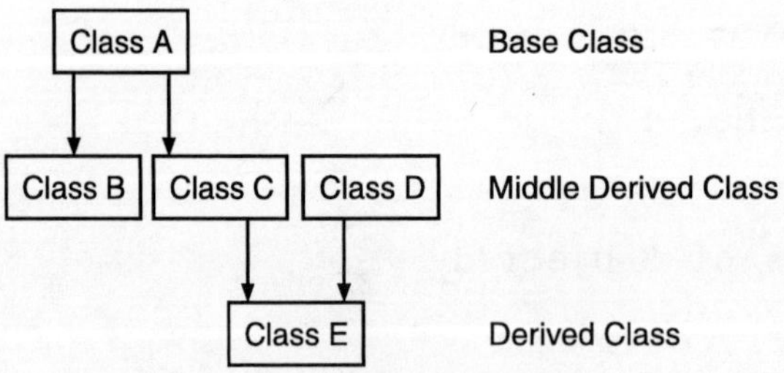

Fig. 5.9

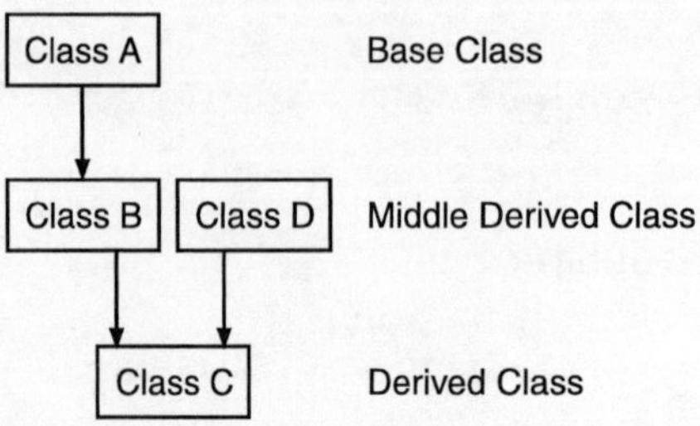

Fig. 5.10

Here we given an example program of student for given figure 5.11 to understand the concept of hybrid inheritance.

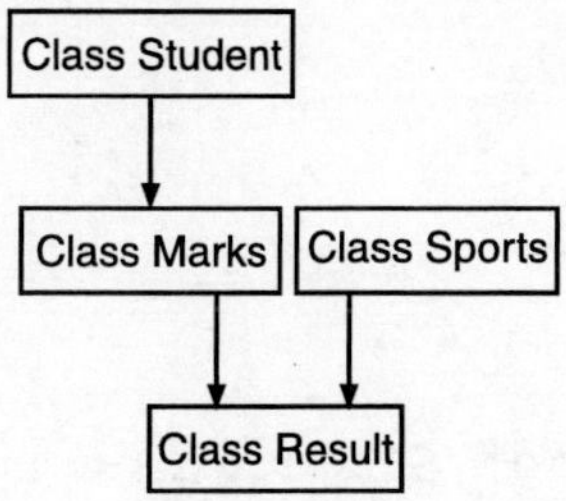

Fig. 5.11

Example 5.14 **Hybrid inheritance**

```
#include<iostream.h>
#include<string.h>
class Student
{
    char name[10];
    int rollno;
public:
    void getdata()
    {
        cout<<"\nEnter the name :"<<endl;
        cin>> name;
        cout<< "\nEnter the Roll No. :"<<endl;
        cin>>rollno;
    }
    void putdata()
    {
        cout<<"\n Name :"<< name;
        cout<<"\n Roll No. :"<<rollno;
    }
};

class Marks : public Student
{
    int marks[4];
protected:
    int total;
public:
    void getmarks();
    void putmarks();
};

void Marks :: getmarks()
{
    for(int i = 0; i <= 4 ; i++)
    {
```

```
            cout<<"\nEnter the Marks of Subject "<< i+1<< ": =" <<endl;
            cin>> marks[i];
        }
    }
    void Marks::putmarks()
    {
        total = 0;
        for(int i = 0; i <= 4 ; i++)
        {
            cout<<"\nMarks of Subject "<< i+1<< ": =" << marks[i];
            total = total + marks[i];
        }
        cout<<"\n Total Marks :"<<total;
    }

    class Sports
    {
        int score;
        char rank[20];
    public:
        void getscore();
        void putscore();
    };

    void Sports :: getscore()
    {
        cout<<"\nEnter the score of sports:"<<endl;
        cin>>score;
    }
    void Sports :: putscore()
    {
            if(score >= 600)
                strcpy(rank, "First");
            elseif(perc >= 450)
                strcpy(rank, "Second");
            elseif(perc >= 300)
                strcpy(rank, "Third");
            else
                strcpy(rank, "Not Satsfatory");

        cout<<" \n Score :"<<score;
        cout<<"\n Rank :"<<rank;
    }
    class Result : public Sports, public Marks
    {
        int perc;
        boolean pass;          // it will store boolean value TRUE for
                               pass and FALSE for fail
        char division[10];
```

```
public:
    void check();
    void display();
};

void Result :: ckeck()
{
    pass = TRUE;
    for(int i = 0; i <= 4 ; i++)
    {
        if (marks[i]<30)
            pass = FALSE;
    }
    perc = (total * 100) / 400;
}

void Result :: display()
{
    if (pass = = TRUE)
    {
        if(perc >= 60)
            strcpy(division, "First");
        elseif(perc >= 45)
            strcpy(division, "Second");
        elseif(perc >= 30)
            strcpy(division, "Third");
        else
            strcpy(division, "Fail");
    }
    else
    {
        cout<<"\nStudent is Fail";
        strcpy(division, "Fail");
        break;
    }

    cout<<" \n Percentage :"<<perc<< "%";
    cout<<"\n Division :"<<division;
}
void main()
{
    Result R;

    R.getdata();
    R.getmarks();
    R.getscore();

    R.putdata();
    R.putmarks();
```

```
    R.putscore();
    S.check();
    S.display();
}
```

OUTPUT

```
Enter the Name :
Ravi

Enter the Roll No. :
1104

Enter the Marks of Subject 1
65

Enter the Marks of Subject 2
69

Enter the Marks of Subject 3
76

Enter the Marks of Subject 4
54

Enter the score of Sports :
540

Name : Ravi

Roll No.: 1104

Marks of Subject 1
65

Marks of Subject 2
69

Marks of Subject 3
76

Marks of Subject 4
54

Score : 540

Rank : First

Total Marks : 264
```

```
Percentage : 66 %

Division : First
```

Related Questions In Exams (Most Important)

Q.1. What does the inheritance means in C++? [May-2007]

Explain Inheritance with example. [Dec-2007]

What is Inheritance? What are various types of inheritance? [Dec-2008]

What is the usefulness of Inheritance? What are various types of inheritance? [May-2009]

Ans. Refer Section 5.1 and 5.4.

Q.2. What is the difference between public, private & protected inheritance? [Dec-2007]

Discuss the effect of inheritance on the visibility of members in :

(a) Public derivation

(b) Private derivation

(c) Protected derivation

Ans. Refer Section 5.3.

5.5 PROTECTED MEMBERS

Although the private members of a base class are inherited by a derived class, they are not accessible to it. See an example :

Example 5.15 **Protected Member**

```
#include<iostream.h>

class First
{
protected:
    int a;              // a is protected variable will inherit in
                        derived classes
public:
    void getdata()
    {
        cout<<"\nEnter the value of a :" <<endl;
        cin>> a;
    }
    void putdata()
    {
        cout<<"\nThe value of a is :"<< a;
    }
};

class Second : public First
{
protected:
    int b;              // b is protected variable will inherit in
                        derived classes
```

```
public:
    void indata()
    {
        cout<<"\nEnter the value of b :" <<endl;
        cin>> b;
    }
    void outdata()
    {
        cout<<"\nThe value of b is :"<< b;
    }

};

class Third : public Second
{
    int c;
public:
    void input()
    {
        cout<<"\nEnter the value of c :" <<endl;
        cin>> c;
    }
    void output()
    {
        cout<<"\nThe value of c is :"<< c;
    }

    void addition ()
    {
        int d = a + b + c;
        cout<<"\n Addition of variables: "<< d;
    }
};

void main()
{

    Third T ;

    T.getdata();      // member function of class First

    T.indata();       // member function of class Second

    T.input()         // member function of class Third

    T.putdata();
    T.outdata();
    T.output();

    T.addition();
}
```

OUTPUT

```
Enter the value of a :
2

Enter the value of b :
4

Enter the value of c :
6

The value of a is : 2

The value of b is : 4

The value of b is : 6

Addition of both variables : 12
```

In above example, Third class inherits all the data (private, public or protected) members of First and Second, but is not allowed to directly refer to the private members of First or Second. The idea is that private members should be completely hidden so that they cannot be tampered with by the class clients. This restriction may prove too prohibitive for classes from which other classes are likely to be derived.

The restriction can be relaxed by defining the base class private members as protected instead. As far as the clients of a class are concerned, a protected member is the same as a private member: it cannot be accessed by the class clients. However, a protected base class member can be accessed by any class derived from it.

5.6 CASTING

Suppose we have defined two classes for, respectively, representing lists of options and bitmapped windows :

```
class OptionList
{
public:
OptionList (int n);
~OptionList (void);
//...
};
class Window
{
public:
Window (Rect &bounds);
~Window (void);
//...
};
```

A menu is a list of options displayed within its own window. It therefore makes sense to define Menu by deriving it from Option List and Window :

```
class Menu : public OptionList, public Window {
public:
Menu (int n, Rect &bounds);
~Menu (void);
//...
};
```

For any derived class there is an implicit type conversion from the derived class to any of its *public* base classes. This can be used for converting a derived class object to a base class object, be it a proper object, a reference, or a pointer :

```
Menu menu(n, bounds);
Window win = menu;
Window &wRef = menu;
Window *wPtr = &menu;
```

Such conversions are safe because the derived class object always contains all of its base class objects. The first assignment, for example, causes the Window component of menu to be assigned to win. By contrast, there is no implicit conversion from a base class to a derived class. The reason being that such a conversion is potentially dangerous due to the fact that the derived class object may have data members not present in the base class object. The extra data members will therefore end up with unpredictable values. All such conversions must be explicitly cast to confirm the programmer's intention :

```
Menu &mRef = (Menu&) win; // caution!
Menu *mPtr = (Menu*) &win; // caution!
```

A base class object cannot be assigned to a derived class object unless there is a type conversion constructor in the derived class defined for this purpose. For example, given

```
class Menu : public OptionList, public Window {
public:
//...
Menu (Window&);
};
```

the following would be valid and would use the constructor to convert win to a Menu object before assigning :

```
menu = win; // invokes Menu::Menu(Window&)
```

5.7 OVERRIDING

You can use member function in a derived class that have same name as those in the base class. You might want to do this so that calls in your program work the same way for objects of both base and derived classes.

For example :

Example 5.16 **Function Overriding**

```
#include<iostream.h>
class A
{
```

```
    int a1, a2;
public:
    void getdata()
    {
        cout<<"\nEnter the value of a1 and a2 :"<<endl;
        cin>> a1 >> a2;
    }
    void putdata()
    {
        cout<<"\nThe value of a1 is :"<< a1 << "and a2 :"<<a2;
    }
};

class B : public A
{
    int b1, b2;
public:
    void getdata()
    {
        cout<<"\nEnter the value of b1 and b2 :" <<endl;
        cin>> b1 >> b2;
    }
    void putdata()
    {
        cout<<"\nThe value of b1 is :"<< b1 << "and b2 :"<<b2;
    }
};

void main()
{

        B Ob1;

        Ob1.A :: getdata();      //function overriding, member
                                 function of class A
        Ob1.getdata();           // member function of class B

        Ob1.A :: putdata();      //function overriding, member
                                 function of class A
        Ob1.putdata();           / member function of class B

}
```

OUTPUT

```
Enter the value of a1 and a2 :
2
3
Enter the value of b1 and b2 :
4
5
```

```
The value of a1 is : 2 and a2 is : 3

The value of b1 is : 4 and b2 is : 5
```

Example 5.17 **Function overriding**

```
#include<iostream.h>

class First
{
protected:
    int a;      // a is protected variable will inherit in derived
                classes
public:
    void getdata()
    {
        cout<<"\nEnter the value of a :" <<endl;
        cin>> a;
    }
    void putdata()
    {
        cout<<"\nThe value of a is :"<< a;
    }
};

class Second : public First
{
    int b;
public:
    void getdata()
    {
        First :: getdata();     //function overriding
        cout<<"\nEnter the value of b :" <<endl;
        cin>> b;
    }
    void putdata()
    {
        First :: putdata();     //function overriding
        cout<<"\nThe value of b is :"<< b;
    }
    void addition ()
    {
        int d = a + b ;
        cout<<"\n Addition of variables: "<< d;
    }
};

void main()
{
```

```
        Second S;
        S.getdata();                   // member function of class Second
        S.putdata();
        S.addition();
    }
```

OUTPUT

```
Enter the value of a :
2

Enter the value of b :
4

The value of a is : 2

The value of b is : 4

Addition of variables : 6
```

Related Questions In Exams

Q.1. Differentiate between the Overloading & Overriding? *[Dec-2007]*

Ans. Differences :

Overloading	Overriding
1. All function in same class.	1. all functions in different classes.
2. All functions have different signature.	2. all functions have same signature
3. Can be define without classes.	3. must be define in classes.
4. Can be access by directly or (.) dot with objects.	4. must be call with (::) scope resolution operator.

Q.2. Write short notes on Function overriding.

Ans. Refer Section 5.7.

5.8 CONSTRUCTOR AND DESTRUCTORS

A derived class may have constructors and a destructor. Since a derived class may provide data members on top of those of its base class, the role of the constructor and destructor is to, respectively, initialise and destroy these additional members. When an object of a derived class is created, the base class constructor is applied to it first, followed by the derived class constructor. When the object is destroyed, the destructor of the derived class is applied first, followed by the base class destructor. In other words, constructors are applied in order of derivation and destructors are applied in the reverse order. For example, consider a class C derived from B which is in turn derived from A. Figure 5.12 illustrates how an object c of type C is created and destroyed.

```
class A { /*... */ }
class B : public A { /*... */ }
class C : public B { /*... */ }
```

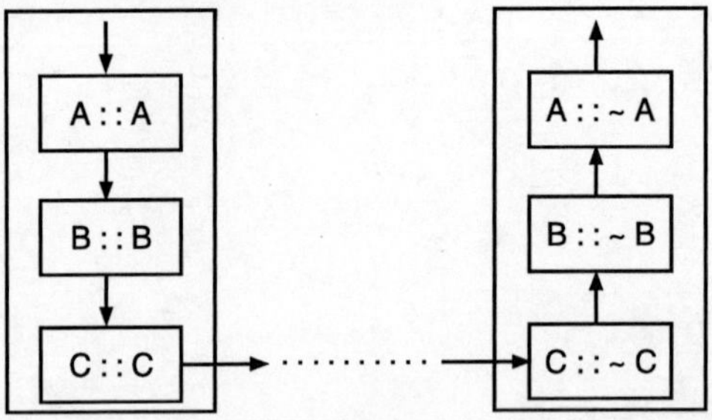

Fig. 5.12 Derived class object construction and destruction order

The constructor of a derived class whose base class constructor requires arguments should specify these in the definition of its constructor. To do this, the derived class constructor explicitly invokes the base class constructor in its member initialisation list.

For example, the Second constructor passes its argument to the First class constructor in this way :

```
Second(int x,int y) :First(x)
{
b = y;
}
```

Example 5.18 **Constructor in inheritance**

```
#include<iostream.h>

class First
{
    int a;
public:
    First(int x) { a = x; }

    void putdata()
    {
        cout<<"\nThe value of a is :"<< a;
    }
};

class Second : public First
{
    int b;
public:
    Second(int x,int y) :First(x)     // calling base class
                                      constructor { b = y; }
```

```
        void putdata()
        {
            First :: putdata();           //function overriding
            cout<<"\nThe value of b is :"<< b;
        }

    };

    void main()
    {

        Second S(10,20);

        S.putdata();

    }
```

OUTPUT

```
    The value of a is : 10

    The value of b is : 20
```

5.9 COMPOSITION

As you know class may have collection of different data types as data member, therefore it may have objects of other classes as data member. When any class contains an object of another class as a member, this class is called a *container class*. See example :

```
class X
{
    :
    :
};

class Y
{
    :
    :
public :
    X   obj;
};
```

In contrast, a class Y that has a member of another class X is said to have X or Y contain X. This relationship is called a *membership* or a "has a" relationship.

Container classes generally come in two different varieties.

- **Value Containers** are compositions that store copies of the objects that they are holding (and thus are responsible for creating and destroying those copies).

- **Reference Containers** are aggregations that store pointers or references to other objects (and thus are not responsible for creation or destruction of those objects).

Definition : When any class contains one or more objects of another classes as member, this class is called a *container class* that may be refer as composition or aggregation.

In real-life, complex objects are often built from smaller, simpler objects. For example, a car is built using a metal frame, an engine, some tires, a transmission, a steering wheel, and a large number of other parts. A personal computer is built from a CPU, a motherboard, some memory, etc. Even you are built from smaller parts: you have a head, a body, some legs, arms, and so on. This process of building complex objects from simpler ones is called ***composition*** (also known as object composition).

More specifically, composition is used for objects that have a has-a relationship to each other. A car has-a metal frame, has-an engine, and has-a transmission. A personal computer has-a CPU, a motherboard, and other components. You have-a head, a body, some limbs.

5.9.1 When to use Composition

Consider the example of Jet_plane class. You might want to consider deriving a class for Jet_plane class from a class called *engine*. However a Jet_plane is not an engine but has an engine. To be able to decide whether to use inheritance or membership you can ask whether a Jet_plane has more than one engine. If that is possibility, it is most likely that the relationship will be "has a" rather than "is a" relationship. Not all relationship will fall under clear-cut categories. Therefore using a derived class or a contain class is a matter to be considered at the design stage. Let see an example in section 3.9 of chapter-3.

5.9.2 Pre-defined Container Classes

C++ also contain pre-defined container classes

A container class is a class that is used to hold objects in memory or external storage. A container class acts as a generic holder. A container class has a pre-defined behaviour and a well-known interface. A container class is a supporting class whose purpose is to hide the topology used for maintaining the list of objects in memory. When a container class contains a group of mixed objects, the container is called a *heterogeneous container*; when the container is holding a group of objects that are all the same, the container is called a homogeneous container.

A container class is defined as a class that gives you the power to store any type of data. There are two types of container classes available in STL in C++, namely "Simple Container Classes" and "Associative Container Classes". An Associative Container class associates a key to each object to minimise the average access time.

Simple Container Classes

- vector<>
- lists<>
- stack<>
- queue<>
- deque<>

Associative Container Classes

- map<>
- set<>

- multimap<>
- multiset<>

Example 5.19 **Vector container class**

```
#include <iostream>
#include <vector>
#include <conio.h>
using namespace std;
int main()
{
vector<char> codes;
codes.assign(10);      //Assigning the size to 10.
cout<<codes.size();    //prints the size of the vector
getch();
return 0;
}
```

The next overloaded match takes 2 pointer as arguments. Here is a sample code.

Example 5.20 **Vector container class**

```
#include <iostream>
#include <vector>
#include <conio.h>
using namespace std;

int main()
{
char a[3]={'a','b','c'};
vector<char> orders;
orders.assign(a,a+3);
cout<<orders.size();
getch();
return 0;
}
```

Types of Container Classes

There are two types of container classes on the basis of containing objects.

1. Homogeneous container
2. Heterogeneous container

Homogeneous Container : When a container class contains a group of objects of a single class is called homogeneous container classes.

Heterogeneous Container : When a container class contains a group of mixed objects of different-different classes is called heterogeneous container classes.

Related Questions In Exams

Q.1. What is containership? How does it differ from inheritance? Explain. *[May-2007]*

Differentiate between Composition & Inheritance? *[Dec-2007]*

Compare composition to inheritance. *[May-2008]*

Explain with the help of example the difference between containership and inheritance. *[Dec-2008]*

Ans. Refer Section 5.9.

Differences :

Inheritance	*Containership/Composition*
1. It is "is a" relationship.	1. It is "has a" relationship.
2. It contains parent and child classes.	2. No need of parent and child classes.
3. In this copies all inherited functions to derived class.	3. Not copied only accessing by object.
4. It is done by only the classes.	4. It is done by classes and objects.
5. It is faster process at runtime.	5. It is slower process at runtime.

Summary

This chapter is fully dedicated to inheritance. In the C++ inheritance is the best feature of sharing the resources among the classes.

With help of example program, we explained the implementation of the inheritance. We have given inheritance structure in the figures to understand in a better way.

Same other concept related to inheritance are also defined such as virtual classes, Abstract classes and function overriding.

Solved Problems

Q. 5.1. Is inheritance important to C++?

Ans. Yes. Inheritance is what separates abstract data type (ADT) programming from OO programming.

Q. 5.2. When would we use inheritance?

Ans. As a specification device.

Human beings abstract things on two-dimensions: part-of and kind-of. A Ford Taurus is-a-kind-of-a Car, and a Ford Taurus has-a Engine, Tires, etc. The part-of hierarchy has been a part of software since the ADT style became relevant; inheritance adds "the other" major dimension of decomposition.

Q. 5.3. Why derived class can't access private things from base class?

Ans. To protect you from future changes to the base class.

Derived classes do not get access to private members of a base class. This effectively "seals off" the derived class from any changes made to the private members of the base class.

Q. 5.4. Are inherited members and functions passed along to subsequent generations? If Dog class derives from Mammal class, and Mammal derives from Animal class, does Dog inherit Animal's functions and data?

Ans. Yes. As derivation continues, derived classes can inherit the sum of all the functions and data in all their base classes.

Q. 5.5. Can a derived class make a public base function private?

Ans. Yes, and it remains private for all subsequent derivation.

Q. 5.6. Why is switching on the runtime type of an object bad?

Ans. With large programs, the switch statements become big and hard to maintain. The point of virtual functions is to let the virtual table, rather than the programmer, determine the runtime type of the object.

Q. 5.7. Why bother making an Abstract Data Type—why not just make it non-abstract and avoid creating any objects of that type?

Ans. The purpose of many of the conventions in C++ is to enlist the compiler in finding bugs, so as to avoid runtime bugs in code that you give your customers. Making a class abstract, that is, giving it pure virtual functions, causes the compiler to flag any objects created of that abstract type as errors.

Exercise

Q. 5.1. What is Base and Drived classes ?

Q. 5.2. What are the mode of Inheritance ?

Q. 5.3. What are the types of Inheritance ?

Q. 5.4. How protected member works and compare with private members ?

Q. 5.5. What is function overriding and how differ from overloading ?

Q. 5.6. How constructurs and destructors are works in inheritance ?

❑❑❑

6

POLYMORPHISM

CHAPTER OUTLINE

6.1 INTRODUCTION

The object oriented programming introduces an important feature polymorphism. The polymorphism a mechanism to use a single name with multiple forms(such as function overloading). Because polymorphism means one name many forms.

There are two type of polymorphism :

1. Static Binding or Compile Time Polymorphism
2. Dynamic Binding or Run Time Polymorphism

6.2 STATIC BINDING

This is a form of polymorphism, in this form the selection of the function invocation is decided at the time of compilation. That's why we have also say that, this is an early binding or compile time polymorphism. Early binding means simply that an object is bound to its function call at compile time. The static binding can be achieve by two ways

(*i*) Function Overloading
(*ii*) Operator Overloading

6.2.1 Function Overloading

The function overloading is used to define distinct functions which share the same name, each of which has a unique signature. When an overloaded function is called, the compiler decides which version of the function to use, that's why function overloading is comes under compile time polymorphism *i.e.,* static binding.

To learn more about the function overloading refer section 3.12

6.2.2 Function Overriding

The function overriding is used to define distinct classes which share the same function name and same signature, but each of which has a unique definition.

When overriding function is called, the compiler decides which definition of function and from which class to use, that's why function overriding is comes under compile time polymorphism *i.e.,* static binding.

To learn more about the function overriding refer section 5.7.

6.3 DYNAMIC BINDING

If the member function is selected while program is running then this is called dynamic binding or runtime polymorphism. In the runtime polymorphism the function link with the class very late (means after the compilation process) thus this polymorphism is also know as late binding. The dynamic or late binding can be achieve by the concept of virtual functions.

Polymorphism in OOP means One name, Multiple Forms or One Item, Multiple Tasks.

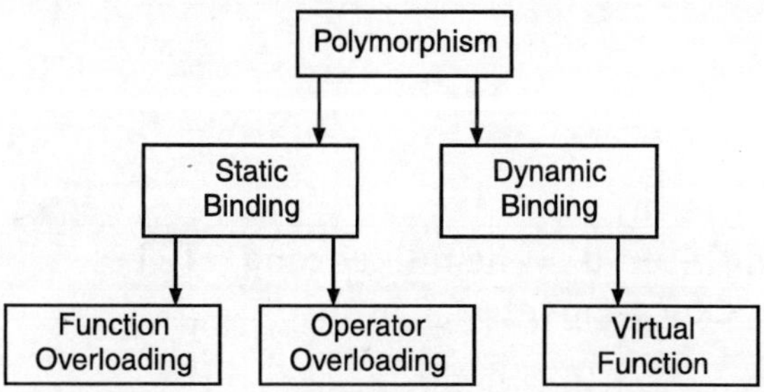

Fig. 6.1

Before discussing on the virtual function, we just discussing on the pointers with classes and objects.

6.4 POINTERS WITH CLASSES

C++ supports dynamic binding which is one of the powerful features of object oriented programming. To achive dynamic binding we require virtual functions, which implemented by concept of pointers with the classes and object. Here we disscuss concept of pointer, after that we explain virtual function.

6.4.1 Pointer to Objects

It is perfectly valid to create pointers that point to classes. We simply have to consider that once declared, a class becomes a valid type, so we can use the class name as the type for the pointer. For example :

```
CRectangle * prect;
```

is a pointer to an object of class CRectangle.

As it happened with data structures, in order to refer directly to a member of an object pointed by a pointer we can use the arrow operator (->) of indirection. Here is an example with some possible combinations :

Example 6.1 **Pointer to classes**

```
#include <iostream.h>

class CRectangle
{
    int width, height;
public:
    void set_values (int, int);
    int area (void)
    {
        return (width * height);
    }
};

void CRectangle::set_values (int a, int b)
{
        width = a;
        height = b;
}

int main ()
{
        CRectangle a, *b, *c;
        CRectangle * d = new CRectangle[2];
        b= new CRectangle;
        c= &a;
        a.set_values (1,2);
        b->set_values (3,4);
        d->set_values (5,6);
        d[1].set_values (7,8);
        cout << "a area: " << a.area() << endl;
        cout << "*b area: " << b->area() << endl;
        cout << "*c area: " << c->area() << endl;
        cout << "d[0] area: " << d[0].area() << endl;
        cout << "d[1] area: " << d[1].area() << endl;
        return 0;
        }
```

OUTPUT

```
a area: 2
*b area: 12
*c area: 2
d[0] area: 30
d[1] area: 56
```

Next you have a summary on how can you read some pointer and class operators (*, &,.,->, [ ]) that appear in the previous example :

expression	*can be read as*
*x	pointed by x
&x	address of x
x.y	member y of object x
x->y	member y of object pointed by x
(*x).y	member y of object pointed by x (equivalent to the previous one)
x[0]	first object pointed by x
x[1]	second object pointed by x
x[n]	(n+1)th object pointed by x

6.4.2 Array of Pointers

If the object of a class that contain character strings would not be of the same size. In this cases, we define an array of pointers to objects that can be access the individual objects. For example :

Example 6.2 **Array of pointers**

```
#include<iostream.h>
#include<string.h>
class city
{
    protected:
        int lenth;
        char*name;
    public:
        city()
        {
            lenth = 0;
            name = new char[len + 1];
        }
        void getname()
        {
            char *a;
            a = new char[30];

            cout<< "\nEnter city name :"<<endl;
```

```
            cin>> a;
            len = strlen(a);
            name = new char[len + 1];
            strcpy(name, a);
        }
        void putname()
        {
            cout<< "\nCity Name :"<<name;
        }
};

int main()
{
    city *p[10];

    int n = 1;
    int opt;

    do
    {
        p[n] = new city;
        p[n++]-> getname();

        cout<< "\n Enter 1 for more entry of city name or 0 for
        exit"<<endl;
        cin>>opt;
    }while(opt);

    cout<< "\n\n";
    for(int i=1; i<=n; i++)
        p[i]->putname();
    }
    return 0;
}
```

OUTPUT

```
Enter city name :
Banaras

Enter 1 for more entry of city name or 0 for exit
1
Enter city name :
Mumbai

Enter 1 for more entry of city name or 0 for exit
1
```

```
Enter city name :
Delhi

Enter 1 for more entry of city name or 0 for exit
0

City Name : Banaras
City Name : Mumbai
City Name : Delhi
```

6.4.3 Pointer to Derived Classes

Pointer to objects of base class are type compatible with pointers to objects of a derived class.

For example :

```
class Base
    {
    ........
    };
class Derive : public Base
    {
    ........
    };
Base *P;
Base B;
Derive D;
P = &B;
P = &D
```

In above statement P = &D is also valid because the pointer of Base is type compatible to Derive class.

We will discuss more on pointer to derived classes in next section virtual function.

Related Questions In Exams (Most Important)

Q.1. Differentiate between static and dynamic linking. *[Dec-2006]*

What is polymorphism? What is the advantages of using it? Differentiate between compile time and runtime polymorphism using example. *[Dec-2006]*

Explain polymorphism with example. *[May-2008]*

What is polymorphism? In what ways polymorphism can be achieved? *[Dec-2008]*

Ans. Refer Section 6.1 and 6.2.

Q.2. What is the difference between Static binding and dynamic binding? *[Dec-2008]*

What do you mean by compile time polymorphism and runtime polymorphism? *[May-2009]*

Differentiate between late binding and early binding? Write the appropriate programs to explain the concept of each. *[Dec-2009]*

Ans. Refer Section 6.1 and 6.2.

6.5 VIRTUAL FUNCTIONS

The virtual functions allow derived classes to provide different versions of a base class function. You can use the *virtual* keyword to declare a virtual function in a base class. By declaring the function prototype in the usual way and then prefixing the declaration with the *virtual* keyword. For example :

Example 6.3 **Virtual members**

```
#include <iostream.h>
class Polygon
{
protected:
    int width, height;
public:
    void set_values (int a, int b)
    {
        width=a;
        height=b;
    }
    virtual int area ()
    {
        return (0);
    }
};

class Rectangle: public Polygon
{
public:
    int area ()
    {
        return (width * height);
    }
};

class Triangle: public Polygon
{
public:
    int area ()
    {
        return (width * height / 2);
    }
};

int main ()
```

```
{
        Rectangle rect;
        Triangle trgl;
        Polygon poly;

        Polygon * ppoly1 = &rect;
        Polygon * ppoly2 = &trgl;
        Polygon * ppoly3 = &poly;

        ppoly1->set_values (4,5);
        ppoly2->set_values (4,5);
        ppoly3->set_values (4,5);

        cout << ppoly1->area() << endl;
        cout << ppoly2->area() << endl;
        cout << ppoly3->area() << endl;
        return 0;
}
```

OUTPUT

```
20
10
0
```

Now the three classes (Polygon, Rectangle and Triangle) have all the same members: width, height, set_values() and area( ).

The member function area() has been declared as virtual in the base class because it is later re-defined in each derived class. You can verify if you want that if you remove this virtual keyword from the declaration of area() within Polygon, and then you run the program the result will be 0 for the three polygons instead of 20, 10 and 0. That is because instead of calling the corresponding area() function for each object (Rectangle::area(), Triangle::area() and Polygon::area(), respectively), Polygon::area() will be called in all cases since the calls are *via.,* a pointer whose type is Polygon*.

Therefore, what the virtual keyword does is to allow a member of a derived class with the same name as one in the base class to be appropriately called from a pointer, and more precisely when the type of the pointer is a pointer to the base class but is pointing to an object of the derived class, as in the above example.

A class that declares or inherits a virtual function is called a *polymorphic class.*

Note that despite of its virtuality, we have also been able to declare an object of type Polygon and to call its own area() function, which always returns 0.

6.5.1 Pure Virtual Function

To declare a pure virtual function (which automatically declares an abstract class), prefix the prototype with the virtual keyword, and set the function equal to zero.

```
virtual int funct1(void);        // A virtual function declaration.
virtual int funct2(void) = 0;    // A pure function declaration.
```

An abstract base Polygon class could look like this :

```
// abstract class Polygon
class Polygon {
protected:
    int width, height;
public:
    void set_values (int a, int b)
        { width=a; height=b; }
    virtual int area () =0;
};
```

Pure Virtual Function is a new concept, where function assigned with a value, see :
return_type function_name() = 0;

Notice how we appended =0 to virtual int area () instead of specifying an implementation for the function. This type of function is called a *pure virtual function*, and all classes that contain at least one pure virtual function are *abstract base classes*.

Related Questions In Exams

Q.1. What is a pure virtual function? Justify its use in an example. *[Dec-2006]*
Write a short note on virtual function? *[Dec-2007]*
Write a short note on virtual function? *[Dec-2008]*
Distinguish between the virtual functions and pure virtual functions. Give example. *[May-2009]*

Ans. Refer Section 6.5 and 6.5.1.

6.6 ABSTRACT CLASSES

An **abstract class** is a class which is never used directly but provides a skeleton for other classes to be derived from it. Typically, all the member functions of an abstract are virtual and have dummy implementations. The following is a simple example of an abstract class :

```
class Database
{
public:
virtual void Insert (Key, Data) {}
virtual void Delete (Key) {}
virtual Data Search (Key) {return 0;}
};
```

It provides a skeleton for a database-like classes. Examples of the kind of classes which could be derived from database include: linked-list, binary tree, and B-tree. First derive a B-tree class from Database and then derive a B*-tree from B-tree :

```
class BTree : public Database { /*...*/ };
class BStar : public BTree { /*...*/ };
```

When any class that contain at least one pure virtual function are also be an *abstract base class*.

The main difference between an abstract base class and a regular polymorphic class is that because in abstract base classes at least one of its members lacks implementation we cannot create instances (objects) of it.

But a class that cannot instantiate objects is not totally useless; We can create pointers to it and take advantage of all its polymorphic abilities. Therefore a declaration like:

```
Polygon poly;
```

would not be valid for the abstract base class we have just declared, because tries to instantiate an object. Nevertheless, the following pointers :

```
Polygon * ppoly1;
Polygon * ppoly2;
```

would be perfectly valid.

This is so for as long as Polygon includes a pure virtual function and therefore it's an abstract base class. However, pointers to this abstract base class can be used to point to objects of derived classes.

Here you have the complete example :

Example 6.4 **Abstract base class & pure virtual function**

```
#include <iostream.h>
class Polygon
{
protected:
    int width, height;
public:
    void set_values (int a, int b)
    {
        width=a;
        height=b;
    }
    virtual int area (void) =0;
};

class Rectangle: public Polygon
{
public:
    int area (void)
    {
        return (width * height);
    }
};

class Triangle: public Polygon
{
public:
```

```
        int area (void)
        {
        return (width * height / 2);
        }
    };

    int main ()
    {
        Rectangle rect;
        Triangle trgl;

        Polygon * ppoly1 = &rect;
        Polygon * ppoly2 = &trgl;

        ppoly1->set_values (4,5);
        ppoly2->set_values (4,5);

        cout << ppoly1->area() << endl;
        cout << ppoly2->area() << endl;
        return 0;
    }
```

OUTPUT

```
20
10
```

If you review the program you will notice that we refer to objects of different but related classes using a unique type of pointer (Polygon*). This can be tremendously useful. For example, now we can create a function member of the abstract base class Polygon that is able to print on screen the result of the area() function even though Polygon itself has no implementation for this function :

Example 6.5 **abstract base class & pure virtual function**

```
// pure virtual members can be called
// from the abstract base class
#include <iostream.h>
class Polygon
{
    protected:
        int width, height;
    public:
        void set_values (int a, int b)
        {
            width=a;
            height=b;
```

```
        }
    virtual int area (void) =0;
    void printarea (void)
    {
        cout << this->area() << endl;
    }
};

class Rectangle: public Polygon
{
public:
    int area (void)
    {
        return (width * height);
    }
};

class Triangle: public Polygon
{
public:
    int area (void)
    {
        return (width * height / 2);
    }
};
int main ()
{
    Rectangle rect;
    Triangle trgl;

    Polygon * ppoly1 = &rect;
    Polygon * ppoly2 = &trgl;

    ppoly1->set_values (4,5);
    ppoly2->set_values (4,5);

    ppoly1->printarea();
    ppoly2->printarea();
    return 0;
}
```

OUTPUT

```
20
10
```

Virtual members and abstract classes grant C++ the polymorphic characteristics that make object-oriented programming such a useful instrument in big projects. Of course, we have seen very simple uses of these features, but these features can be applied to arrays of objects or dynamically allocated objects.

Let's end with the same example again, but this time with objects that are dynamically allocated :

Example 6.6. **Dynamic allocation and polymorphism**

```
#include <iostream.h>
class Polygon
{
protected:
    int width, height;
public:
    void set_values (int a, int b)
    {
        width=a;
        height=b;
    }
    virtual int area (void) =0;
    void printarea (void)
    {
        cout << this->area() << endl;
    }
};

class Rectangle: public Polygon
{
public:
    int area (void)
    {
        return (width * height);
    }
};

class Triangle: public Polygon
{
public:
    int area (void)
    {
        return (width * height / 2);
    }
};

int main ()
```

```
{
    Polygon * ppoly1 = new Rectangle;
    Polygon * ppoly2 = new Triangle;
    ppoly1->set_values (4,5);
    ppoly2->set_values (4,5);

    ppoly1->printarea();
    ppoly2->printarea();

    delete ppoly1;
    delete ppoly2;
    return 0;
}
```

OUTPUT

```
20
10
```

Notice that the ppoly pointers :

```
Polygon * ppoly1 = new Rectangle;
Polygon * ppoly2 = new Triangle;
```

are declared being of type pointer to Polygon but the objects dynamically allocated have been declared having the derived class type directly.

6.6.1 Concrete Classes

Concrete classes are those classes which are leaf classes in the class hierarchy. Which always uses the features of their parents classes.

Class is the way to binding the attributes (data structure) and methods (operations) into a single unit, this feature is called *encapsulation*. But in the hierarchy of classes we have created objects for some classes. So, which classes are instantiable, are called ***concrete classes***.

> **Concept**
>
> *A concrete class has its direct instance in the system (object).*

Thus, a class which have a direct instance is called concrete class. You can define any class as a concrete class in the inheritance tree, which is required in your application.

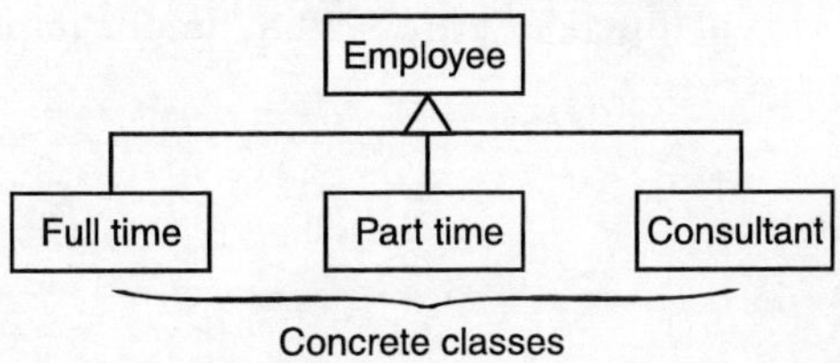

Fig. 6.2. Concrete class

In the figure 6.2 all classes on second level of the inheritance tree are concrete classes. Such as Full time, Part time and Consultant classes are concrete classes, they will contain direct instance in the system.

The concrete class may be a leaf class in the inheritance tree. And only concrete classes may be leaf classes in the inheritance tree. It is not necessary that every leaf class will be concrete class, any root-class of the tree may be a concrete class according to requirement of the application.

Notice that the ppoly pointers:

```
Polygon * ppoly1 = new Rectangle;
Polygon * ppoly2 = new Triangle;
```

are declared being of type pointer to Polygon but the objects dynamically allocated have been declared having the derived class type directly.

Related Questions In Exams

Q.1. What are abstract base classes and concrete classes? *[Dec-2007]*
Ans. Refer Section 6.6 and 6.6.1.

Summary

This chapter explain how C++ implements the concept of polymorphism. Polymorphism supported by C++ both at compile time and at runtime. As discussed, compile time polymorphism is achieved by overloading functions and operators. Runtime polymorphism is accomplished by using inheritance and virtual functions.

Exercise

Q. 6.1. What do you mean by late binding?
Q. 6.2. Give an example of compile time binding?
Q. 6.3. What are the role of pointers in Polymorphism?
Q. 6.4. What are the virtual functions? Why we need it?
Q. 6.5. What is the role of pure vitual functions? How it is differ from normal virtual functions?
Q. 6.6. What are the features of virtual functions?
Q. 6.7. How can we achieve polymorphism at compile time and runtime ?
Q. 6.8. What is the role of this pointer ?
Q. 6.9. When do we make a virtual function "pure"? What are the implications of making a function a pure virtual functions?

❑❑❑

7

FILES AND INPUT/OUTPUT STREAMS

CHAPTER OUTLINE

- Introduction
  - Pointer to Objects
- Introduction
- I/O Stream Classes
- Opening & Closing of File
  - Open a file
  - Closing a file
- Modes of File
- Sequential Access Files
  - Creating & Writing Files
  - Reading Files
- Random Access Files
  - Creating & Writing Files
  - Reading Files
- Stream Manipulators
- Summary

7.1 INTRODUCTION

C++ has no built-in Input/Output (IO) capability. Instead, this capability is provided by a library. The standard C++ IO library is called the ***iostream*** library. The definition of the library classes is divided into three header files. An additional header file defines a set of **manipulators** which act on streams. These are summarised by Table 7.1.

Figure 7.1 relates these header files to a class hierarchy for a UNIX-based implementation of the iostream class hierarchy. The highest-level classes appear unshaded. A user of the iostream library typically works with these classes only. Table 7.2 summarises the role of these high-level classes. The library also provides four pre-defined stream objects for the common use of programs. These are summarised by Table 7.3.

7.2 I/O STREAM CLASSES

A stream may be used for input, output, or both. The act of reading data from an input stream is called ***extraction***. It is performed using the >> operator (called the *extraction operator*) or an iostream member function. Similarly, the act of writing data to an output stream is called ***insertion***, and is performed using the << operator (called the *insertion operator*) or an iostream member function. We therefore speak of 'extracting data from an input stream' and 'inserting data into an output stream'.

Table 7.1 Iostream Header Files.

Header File	Description
iostream.h	Defines a hierarchy of classes for low-level (untyped character level) IO and high-level (typed) IO. This includes the definition of the ios, iotream, ostream, and iostream classes.
fstream.h	Derives a set of classes from those defined in iostream.h for file IO. This includes the definition of the ifstream, ofstream, and fstream classes.
strstream.h	Derives a set of classes from those defined in iostream.h for IO with respect to character arrays. This includes the definition of the istrstream, ostrstream, and strstream classes.
iomanip.h	Defines a set of manipulator which operate on streams to produce useful effects.

Table 7.2 Highest-level Iostream Classes

Form of IO	Input	Output	Input and Output
Standard IO	istream	ostream	iostream
File IO	ifstream	ofstream	fstream
Array of char IO	istrstream	ostrstream	strstream

Table 7.3 Predefined streams

Stream	Type	Buffered	Description
cin	istream	Yes	Connected to standard input (*e.g.*, the keyboard)
cout	ostream	Yes	Connected to standard output (*e.g.*, the monitor)
clog	ostream	Yes	Connected to standard error (*e.g.*, the monitor)
cerr	ostream	No	Connected to standard error (*e.g.*, the monitor)

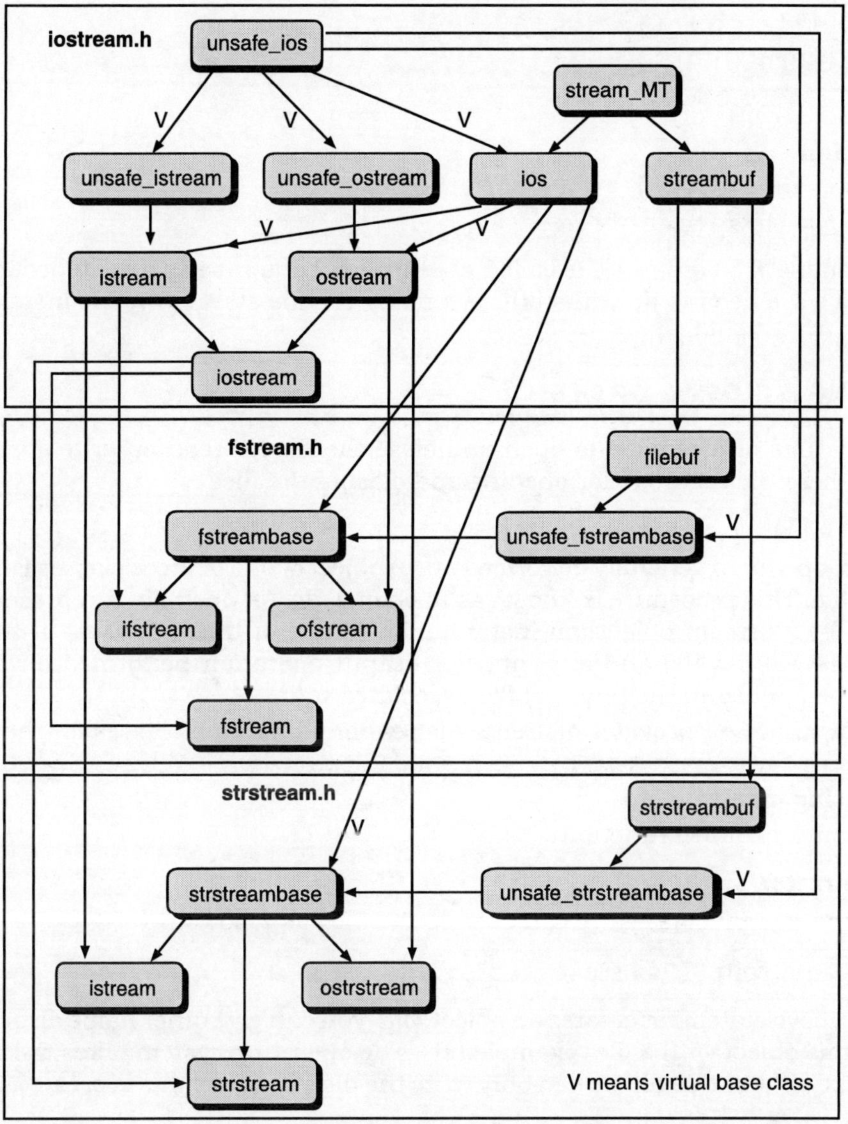

Fig. 7.1 Iostream class hierarchy

Now lets see an example of file :

Example 7.1 A simple file program

```
#include <iostream>
#include <fstream>

int main ()
{
    ofstream myfile;
    myfile.open ("example.txt");
```

```
    myfile << "Writing this to a file.\n";
    myfile.close();
    return 0;
}
```

Execution

```
[file example.txt]
Writing this to a file
```

The example 7.1 creates a file called ***example.txt*** and inserts a sentence into it in the same way we are used to do with cout, but using the file stream ***myfile*** instead.

But let's go step by step :

7.3 OPENING & CLOSING OF FILE

For all programs of files need to open and close the file for read or write operations. We have given here the methods for opening and closing the files.

7.3.1 Open a file

The first operation generally performed on an object of one of these classes is to associate it to a real file. This procedure is known as to *open a file.* An open file is represented within a program by a stream object (an instantiation of one of these classes, in the previous example this was myfile) and any input or output operation performed on this stream object will be applied to the physical file associated to it.

We know that C++ provides ofstream, ifstream and fstream classes for the operations on files. Using these classes we can open a file in two ways :

1. using constructors
2. using member function

Using Constructor

Using the stream classes we can creates the files as follows :

```
ofstream fout ("example.txt");
```

On the above statement creates an object *fout* (you can give other name also) of ofstream class and this object with a file "example.txt". The operating system takes the name of file "example.txt", but in this file we can only write the data because this is created by ofstream class.

A file can be created by ifstream class through constructor as follows :

```
ifstream fin("example.txt");
```

In this statement ifstream class creates an object *fin* with a file "example.txt". in this file we can perform only input operation only(means only read operations).

A file can be created by fstream class through constructor as follows :

```
fstream file("example.txt");
```

In this statement fstream class creates an object *file* with a file "example.txt". In this file we can perform both input and output operation.

Using Member Function

In order to open a file with a stream object we use its member function open() :

```
open (filename, mode);
```

Where filename is a null-terminated character sequence of type const char * (the same type that string literals have) representing the name of the file to be opened, and mode is an optional parameter.

For example

```
ofstream fout;
fout.open ("example.txt");
```

On the above example, first statement creates an object *fout* of ofstream class and in next statement *fout* used a member function "open()" with a file "example.txt". The operating system takes the name of file "example.txt", but in this file we can only write the data because this is created by ofstream class.

A file can be created by ifstream class through its same member fuction as follows :

```
ifstream fin;
fin.open ("example.txt");
```

In this example an object of ifstream class creates a file "example.txt". In this file we can perform only input operation only(means only read operations).

```
fstream file;
file.open ("example.txt");

// file.open ("example.txt", ios::out | ios::in)
```

In this example fstream class creates an object *file* with a file "example.txt". In this file we can perform both input and output operation. We can use the an another method to open file in both modes, which is given in comment. We will discuss about the modes in next sections.

7.3.2 Closing a file

When we are finished with our input and output operations on a file we shall close it so that its resources become available again. In order to do that we have to call the stream's member function close(). This member function takes no parameters, and what it does is to flush the associated buffers and close the file :

```
myfile.close();
```

Once this member function is called, the stream object can be used to open another file, and the file is available again to be opened by other processes.

In case that an object is destructed while still associated with an open file, the destructor automatically calls the member function close().

7.4 MODES OF FILE

We can open a file multiple modes with a combination of the following flags :

ios::in	Open for input operations.
ios::out	Open for output operations.
ios::binary	Open in binary mode.
ios::ate	Set the initial position at the end of the file.If this flag is not set to any value, the initial position is the beginning of the file.
ios::app	All output operations are performed at the end of the file, appending the content to the current content of the file. This flag can only be used in streams open for output-only operations.
ios::trunc	If the file opened for output operations already existed before, its previous content is deleted and replaced by the new one.

All these flags can be combined using the bitwise operator OR (|). For example, if we want to open the file example.bin in binary mode to add data we could do it by the following call to member function open() :

```
ofstream myfile;
myfile.open ("example.bin", ios::out | ios::app | ios::binary);
```

Each one of the open() member functions of the classes ofstream, ifstream and fstream has a default mode that is used if the file is opened without a second argument :

class	**default mode parameter**
ofstream	ios::out
ifstream	ios::in
fstream	ios::in | ios::out

For ifstream and ofstream classes, ios::in and ios::out are automatically and respectively assumed, even if a mode that does not include them is passed as second argument to the open() member function.

The default value is only applied if the function is called without specifying any value for the mode parameter. If the function is called with any value in that parameter the default mode is overridden, not combined.

Files are used to store input-output data for long time on hard-disk.

File streams opened in binary mode, perform input and output operations independently of any format considerations. Non-binary files are known as *text files*, and some translations may occur due to formatting of some special characters (like newline and carriage return characters).

Since the first task that is performed on a file stream object is generally to open a file, these three classes include a constructor that automatically calls the open() member function and has the exact same parameters as this member. Therefor, we could also have declared the previous myfile object and conducted the same opening operation in our previous example by writing :

```
ofstream myfile ("example.bin", ios::out | ios::app | ios::binary);
```

Combining object construction and stream opening in a single statement. Both forms to open a file are valid and equivalent.

To check if a file stream was successful opening a file, you can do it by calling to member is_open() with no arguments. This member function returns a bool value of true in the case that indeed the stream object is associated with an open file, or false otherwise :

```
if (myfile.is_open()) { /* ok, proceed with output */ }
```

Related Questions In Exams (Most Important)

Q.1. Write a short note on File Handling in C++? *[Dec-2008]*

Write a short note on File Handling in C++? *[May-2009]*

Ans. Refer Section 7.1 to 7.3.

Q.2. Differentiate between put() and write() and get() and read()? [Dec-2006]

Ans.

Function	***Description***
put()	Member of ostream.h , used to handle a single character for output to text file.
write()	Member of ostream.h , used to handle a single character for output to binary file.
get()	Member of istream.h , used to handle a single character for input from text file.
read()	Member of istream.h , used to handle a single character for input from binary file.

Q.3. Discuss the following file mode parameters: ios::app, ios::noreplace, ios::trunk, ios::out. *[Dec-2006]*

What is file mode? Discuss the various file mode options available in C++. *[Dec-2007]*

Ans. Refer Section 7.4.

File Mode	***Description***
ios::app	File opening mode, open with append to end of file.
ios::noreplace	File opening mode, open file if already exist.
ios::trunk	File opening mode, open after deleting contents if it exist.
ios::out	File opening mode, open for write only..

7.5 SEQUENTIAL ACCESS FILES

Refers to reading or writing data records in sequential order, that is, one record after the other. To read record 10, for example, you would first need to read records 1 through 9. This differs from *random access*, in which you can read and write records in any order.

Some programming languages and operating systems distinguish between sequential-access data files and random-access data files, allowing you to choose between the two types. Sequential-access files are faster if you always access records in the same order. Random-access files are faster if you need to read or write records in a random order.

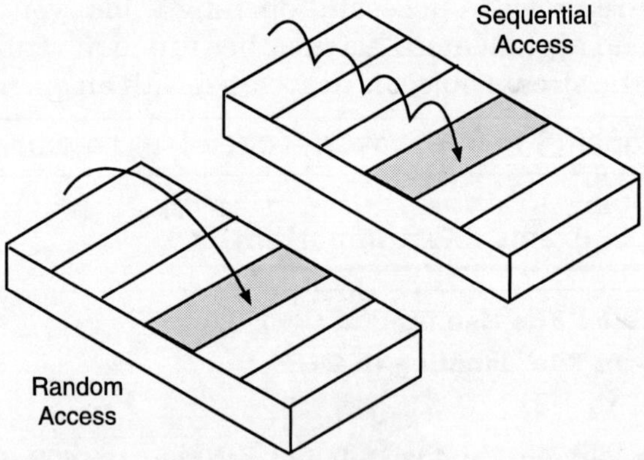

Fig. 7.2 Working of Sequential & Random Access Files

7.5.1 Creating & Writing Files

For creating and writing a sequential file lets see an example :

Example 7.2 **Creating and Writing on Sequential File**

```
#include <iostream>
#include <fstream>
using namespace std;

int main ()
{
    ofstream myfile;
    myfile.open ("example.txt");
    myfile << "Writing this to a file.\n";
    myfile.close();
    return 0;
}
```

Execution

```
[file example.txt]
Writing this to a file
```

Example 7.3 **Creating and Writing on Sequential File with check for opening**

```
#include <iostream>
#include <fstream>
using namespace std;

int main ()
{
    ofstream myfile ("example.txt");
    if (myfile.is_open())
```

```
    {
        myfile << "This is a line.\n";
        myfile << "This is another line.\n";
        myfile.close();
    }
    else cout << "Unable to open file";
    return 0;
}
```

Execution

```
[file example.txt]
This is a line.
This is another line.
```

7.5.2 Reading Files

For reading a sequential file let see an example :

Example 7.4 **Reading a text file**

```
#include <iostream.h>
#include <fstream.h>
#include <string.h>

int main ()
{
    char line[80];
    ifstream myfile ("example.txt");

    if (myfile.is_open())
    {
        while (! myfile.eof() )
        {
            getline (myfile,line);
            cout << line << endl;
        }
    myfile.close();
    }

    else
        cout << "Unable to open file";

    return 0;
}
```

OUTPUT

```
This is a line.
This is another line.
```

Example 7.5 **Reading text file using class**

```
#include <iostream>
#include <fstream>
#include <conio.h>

using namespace std;

class File
{
    ifstream f1;
    ofstream f2;
    char s[200];

    public:
        void ShowFile(char *nume);
        void ReadInFile(char *nume);
};

void File::ReadInFile(char *nume)
{

    f2.open(nume);
    for(int i=0; i<strlen(s); i++)
    {
    f2<<s[i];
    }

    f2.close();
}

void File::ShowFile(char *nume)
{
    f1.open(nume);
    if(f1.fail())
    {
        cout<<"Error opening file!";
        getch();
        exit(1);
    }
    while(!f1.eof())
    {

        for(int i=0; i<strlen(s); i++)
```

```
            {
            f1>>s[i];
            cout<<s[i];
            }
        }
        f1.close();
    }

    int main()
    {
        File F;
        F.ShowFile("slav.txt");
        F.ReadInFile("slav2.txt");

        getch();
        return 0;
    }
```

7.6 RANDOM ACCESS FILES

The terms *random access* and *sequential access* are often used to describe data files. A random-access data file enables you to read or write information anywhere in the file. In a sequential-access file, you can only read and write information sequentially, starting from the beginning of the file.

Both types of files have advantages and disadvantages. If you are always accessing information in the same order, a sequential-access file is faster. If you tend to access information randomly, random access is better.

Random access is sometimes called *direct access.* Operating systems store the files in binary file format. Computers can deal with only binary numbers. But binary files are not readable by humans. Our level of comfort lies only with proper ASCII or UNICODE characters. This article deals with how C++ File I/O class fstream can be used for reading and writing binary files.

7.6.1 Creating & Writing Files

For our C++ File I/O binary file examples, we assume a **struct WebSites** with two members as follows.

```
struct WebSites
{
    char SiteName[100];
    int Rank;
};
```

There are some important points to be noted while doing a write operation.

- The file has to be opened in output and binary mode using the flags ios::out (output mode) and ios::binary(binary mode)
- The function *write* takes two parameters. The first parameter is of type *char* * for the data to be written and the second is of type *int* asking for the size of data to be written to the binary file.

- File has to be closed at the end.

// Sample for C++ File I/O binary file write

```
void write_to_binary_file(WebSites p_Data)
{

    fstream binary_file("c:\\test.dat",ios::out|ios::binary|ios::app);
    binary_file.write(reinterpret_cast<char *>(&p_Data),sizeof
    (WebSites));
    binary_file.close();
}
```

The above C++ File I/O binary sample function writes some data to the function. The file is opened in output and binary mode with *ios::out* and *ios::binary*. There is one more specifier *ios::app*, which tells the Operating system that the file is also opened in *append* mode. This means any new set of data will be appended to the end of file.

Also the *write* function used above, needs the parameter as a character pointer type. So we use a type converter reinterpret_cast to typecast the structure into char* type.

7.6.2 Reading Files

This also has a similar flow of operations as above. The only difference is to open the file using ios::in, which opens the file in read mode.

// Sample for C++ File I/O binary file read

```
void read_from_binary_file()
{
    WebSites p_Data;
    fstream binary_file("c:\\test.dat",ios::binary|ios::in);
    binary_file.read(reinterpret_cast<char*>(&p_Data),sizeof
    (WebSites));
    binary_file.close();

    cout<<p_Data.SiteName<<endl;
    cout<<"Rank :"<< p_Data.Rank<<endl;

}
```

Most of the programs will usually go for ASCII text mode files only. But there will be some occasions where the C++ File I/O with binary files will be very useful.

This " C++ File I/O using binary files" article only gives some very basic samples for read and write. Advanced C++ File I/O operations like seek, checking the file pointer validity etc., are also needed to be learnt while writing bigger programs.

Files may be Text file or Binary file. May be Sequential or Random accessible.

Related Questions In Exams

Q.1. How data is read from a sequential access file? *[Dec-2007]*

How we can create a sequentially access file? *[May-2008]*

How we can create a sequentially access file? *[May-2009]*

What is file? Write a program to create, read and write in a sequential access file. *[Dec-2009]*

Ans. Refer Section 7.5.

Q.2. How data is written to random access file? *[Dec-2006]*

Ans. Refer Section 7.6.1.

Q.3. How we can read a random access files? *[May-2008]*

Ans Refer Section 7.6.2.

Q.4. Compare and contrast between the master file and transaction file. *[Dec-2008]*

Ans. Master File : Master Files are those files which contains the permanent data of the software. These file should not be update directly, software must use the transaction file to update master file. For example: Employee Record.

Transaction File : Transactions files are those files which contain the data used to update the records of master file. These files may contain the duplicate data of master file with updated records.

Q.5. What are unformatted Input/Output? *[May-2008]*

Ans. The files may be two types, text files and binary files. The binary files contains the data in binary format.The binary form of the data is treated as a unformatted data. When we input and output the data from/to file using read() and write() functions, that uses binary data for files. The input/output data is treated as unformatted I/O.

Q.6. Write a program that prompts the user name of text file, reads this file and outputs the number of vowels and number of words in the text. *[Dec-2008]*

Ans.

```
#include<fstream.h>
#include<conio.h>
int main()
{
    clrscr();
    char fname[10];
    cout<<"\nEnter the file name"<<endl;
    cin>>fname;
    fstream f1;
    f1.open(fname,ios::in);
    char c=' ';
    int v=0, w=1;
    while(c!=-1)
    {
        c=f1.get();
        if(c=='i' ||c=='o'||c=='u'||c=='a'||c=='e')v++;
        if(c==' ')w++;
    }
```

```
    f1.close();
    cout<<"\nNumber of Vowels: "<<v;
    cout<<"\nNumber of Words: "<<w;
    return 0;
}
```

Q.7. Write a program to copy the contents of one file to another file byte by byte. *[May-2009]*

Write a program that emulate the DOS COPY command, i.e., it should copy the contents of a text file (such as any CPP file) to another file. *[Dec-2009]*

Ans.

```
#include<fstream.h>
#include<stdlib.h>
#include<conio.h>
int main(int argc, char* argv[])
{
clrscr();
fstream f1,f2;
if(argc!=3)
{
    cout<<"\n Please enter correct arguments & Syntax:\n copy
<file_name1> <file_name2>"<<endl;
    exit(1);
}

f1.open(argv[1],ios::in);
if(f1.fail())
{
    cout<<"\n File :"<<argv[1]<<" does not exist or error in
    opening."<<endl;
    exit(1);
}
f2.open(argv[2],ios::in);
if(!f2.fail())
{
    cout<<"\n File :"<<argv[2]<<" already exist do you want to over
    write (y/n) ?: ";
    char ch;
    cin>>ch;
    if(ch=='n'||ch=='N')
    {
        f1.close();
        f2.close();
        exit(1);
    }
```

```
        f2.close();
    }
    f2.open(argv[2],ios::out);
    char c=' ';
    while(c!=-1)
    {
        c=f1.get();
        f2.put(c);
    }
    f1.close();
    f2.close();
    cout<<"1 File has been copied"<<endl;
    return 0;
    }
```

Q.8. Write a program to print your program code as output. ***[Dec-2009]***

Ans.

```
    #include<fstream.h>
    #include<conio.h>
    int main()
    {
    clrscr();
        char fname[10];
        cout<<"\nEnter your program name"<<endl;
        cin>>fname;
        fstream f1;
        f1.open(fname,ios::in);
        char c=' ';
        while(c!=-1)
        {
            c=f1.get();
            cout<<c;
        }
        f1.close();
        return 0;
    }
```

7.7 STREAM MANIPULATORS

Various stream manipulators can be used to specify the kinds of formatting to be performed during stream-I/O operations. Stream manipulators control the output's format settings. Fig. 7.4 lists each stream manipulator that controls a given stream's format state. All these manipulators belong to class ios_base. We show examples of most of these stream manipulators in the next several sections.

Table 7.4 Format State Stream Manipulators from <iostream>.

Stream Manipulator	Description
skipws	Skip white-space characters on an input stream. This setting is reset with stream manipulator noskipws.
left	Left justify output in a field. Padding characters appear to the right if necessary.
right	Right justify output in a field. Padding characters appear to the left if necessary.
internal	Indicate that a number's sign should be left justified in a field and a number's magnitude should be right justified in that same field (*i.e.*, padding characters appear between the sign and the number).
dec	Specify that integers should be treated as decimal (base 10) values.
oct	Specify that integers should be treated as octal (base 8) values.
hex	Specify that integers should be treated as hexadecimal (base 16) values.
showbase	Specify that the base of a number is to be output ahead of the number (a leading 0 for octals; a leading 0x or 0X for hexadecimals). This setting is reset with stream manipulator no showbase.
showpoint	Specify that floating-point numbers should be output with a decimal point. This is used normally with fixed to guarantee a certain number of digits to the right of the decimal point, even if they are zeros. This setting is reset with stream manipulator noshowpoint.
uppercase	Specify that uppercase letters (*i.e.*, X and A through F) should be used in a hexadecimal integer and that uppercase E should be used when representing a floating-point value in scientific notation. This setting is reset with stream manipulator no uppercase.
showpos	Specify that positive numbers should be preceded by a plus sign (+). This setting is reset with stream manipulator no showpos.
scientific	Specify output of a floating-point value in scientific notation.
fixed	Specify output of a floating-point value in fixed-point notation with a specific number of digits to the right of the decimal point.

Example 7.6 **Using stream manipulator showbase.**

```
#include <iostream>
using std::cout;
using std::endl;
using std::hex;
using std::oct;
using std::showbase;

    int main()
    {
    int x = 100;
```

```
// use showbase to show number base
cout << "Printing integers preceded by their base:" << endl<<
showbase;

cout << x << endl; // print decimal value
cout << oct << x << endl; // print octal value
cout << hex << x << endl; // print hexadecimal value
return 0;
} // end main
```

OUTPUT

Printing integers preceded by their base :

```
100
0 144
0x 64
```

Example 7.7 **Using member function fill and stream manipulator setfill to change the padding character for fields larger than the printed value.**

```
#include <iostream>
using std::cout;
using std::dec;
using std::endl;
using std::hex;
using std::internal;
using std::left;
using std::right;
using std::showbase;

#include <iomanip>
using std::setfill;
using std::setw;

int main()
{
int x = 10000;

// display x
cout << x << " printed as int right and left justified\n"
<< "and as hex with internal justification.\n"
<< "Using the default pad character (space):" << endl;

// display x with base
cout << showbase << setw( 10 ) << x << endl;
```

```
// display x with left justification
cout << left << setw( 10 ) << x << endl;

// display x as hex with internal justification
cout << internal << setw( 10 ) << hex << x << endl << endl;

cout << "Using various padding characters:" << endl;

// display x using padded characters (right justification)
cout << right;
cout.fill( '*' );
cout << setw( 10 ) << dec << x << endl;

// display x using padded characters (left justification)
cout << left << setw( 10 ) << setfill( '%' ) << x << endl;

// display x using padded characters (internal justification)
cout << internal << setw( 10 ) << setfill( '^' ) << hex
<< x << endl;
return 0;
} // end main
```

OUTPUT

```
10000 printed as int right and left justified
and as hex with internal justification.
Using the default pad character (space):
     10000
10000
0x   2710

Using various padding characters :
*****10000
10000%%%%%
0x^^^^2710
```

Example 7.8 **Stream manipulator uppercase.**

```
#include <iostream>
using std::cout;
using std::endl;
using std::hex;
using std::showbase;
using std::uppercase;

int main()
```

```
{
cout << "Printing uppercase letters in scientific" << endl
<< "notation exponents and hexadecimal values:" << endl;

// use std:uppercase to display uppercase letters; use std::hex and
// std::showbase to display hexadecimal value and its base
cout << uppercase << 4.345e10 << endl
<< hex << showbase << 123456789 << endl;
return 0;
} // end main
```

OUTPUT

```
Printing uppercase letters in scientific
notation exponents and hexadecimal values:
4.345E+010
0X75BCD15
```

Example 7.9 **Stream Format States**

```
#include <iostream>
using std::cout;
using std::endl;
using std::ios_base;
using std::oct;
using std::scientific;
using std::showbase;

int main()
{
int integerValue = 1000;
double doubleValue = 0.0947628;

// display flags value, int and double values (original format)
cout << "The value of the flags variable is: " << cout.flags()
<< "\nPrint int and double in original format:\n"
<< integerValue << '\t' << doubleValue << endl << endl;
// use cout flags function to save original format
ios_base::fmtflags originalFormat = cout.flags();
cout << showbase << oct << scientific; // change format

// display flags value, int and double values (new format)
cout << "The value of the flags variable is: " << cout.flags()
<< "\nPrint int and double in a new format:\n"
<< integerValue << '\t' << doubleValue << endl << endl;

cout.flags( originalFormat ); // restore format
```

```
// display flags value, int and double values (original format)
cout << "The restored value of the flags variable is: "
<< cout.flags()
<< "\nPrint values in original format again:\n"
<< integerValue << '\t' << doubleValue << endl;
return 0;
} // end main
```

OUTPUT

```
The value of the flags variable is: 513
Print int and double in original format:
1000 0.0947628

The value of the flags variable is: 012011
Print int and double in a new format:
01750 9.476280e-002
The restored value of the flags variable is: 513
Print values in original format again:
1000 0.0947628
```

Related Questions In Exams

Q.1. What is a stream manipulator? Discuss the following manipulators: setw(), setprecision(), setfill(), setiosflags(). *[May-2007]*

Q.2. What is a stream manipulators? *[Dec-2007]*

Ans. Refer Section 7.7.

Manipulator	*Description*
setw(int w)	Set the field width to w on console.. Example: cout<<setw(10)<<123; Output: _ _ _ _ _ _ _ 123
setprecision(int d)	Set the floating point precision to d. Example: cout<<setw(10)<<setprecision(2)<<123.1; Output: _ _ _ _ _ 123.10
setfill(int c)	Set the fill character to c.
setiosflags()	Set the format flag f.

Summary

This chapter described the standard C++ file I/O handling. To store the information permanently on the secondary storage devices such as hard disk, is common requirement of the applications. So this chapter explains all the file stream classes to do this. The file stream hierarchy shows the concept of derivation of classes and their functions.

In this chapter we have explain the methods of opening and closing file as well as accession method of data from files.

There are some file opening modes, which helps to open the file in various modes.

At last we have given a method of command line arguments programs.

Exercise

Q. 7.1. Why we need files?

Q. 7.2. What are the file stream classes? Explain.

Q. 7.3. Write down the method of opening and closing files.

Q. 7.4. Explain the file opening modes.

Q. 7.5. What are the functions available in C++ to handle the errors for files?

Q. 7.6. How Binary files are differ from others?

Q. 7.7. What do you mean by command line arguments?

Q. 7.8. Explain the arguments of function main().

Q. 7.9. Explain the following functions:

(*i*) tellg() (*ii*) tellp() (*iii*) seekg()

(*iv*) seekp() (*v*) eof() (*vi*) fail()

❑❑❑

8

TEMPLATES AND EXCEPTION HANDLING

CHAPTER OUTLINE

- Introduction to Templates
- Function Templates
- Class Templates
- Templates and Inheritance
- Templates and Friends
- Templates and Static Members
- Exception Handling
- Basics of Exception Handling
- Re-Throwing an Exception
- Exception Specifications
- Processing Unexpected Exception
- Stack Unwinding
- Exception and Inheritance

  Summary

8.1 INTRODUCTION TO TEMPLATES

Templates facilitate the *generic* definition of functions and classes so that they are not tied to specific implementation types. They are invaluable in that they dispense with the burden of re-defining a function or class so that it will work with yet another data type.

We can define two types of templates.

(*i*) Function Templates

(*ii*) Class Templates

A **function template** defines an **algorithm**. An algorithm is a generic recipe for accomplishing a task, independent of the particular data types used for its implementation. For example, the binary search algorithm operates on a sorted array of items, whose exact type is irrelevant to the algorithm. Binary search can therefore be defined as a function template with a type parameter which denotes the type of the array items. This template then becomes a blueprint for generating executable functions by substituting a concrete type for the type parameter. This process is called ***instantiation*** and its outcome is a conventional function.

A **class template** defines a **parameterised type**. A parameterised type is a data type defined in terms of other data types, one or more of which are unspecified. Most data types can be defined independently of the concrete data types used in their implementation. For example, the stack data type involves a set of items whose exact type is irrelevant to the concept of stack. Stack can therefore be defined as a class template with a type parameter which specifies the type of the items to be stored on the stack. This template can then be instantiated, by substituting a concrete type for the type parameter, to generate executable stack classes.

Templates provide direct support for writing reusable code. This in turn makes them an ideal tool for defining generic libraries.

8.2 FUNCTION TEMPLATES

Function templates are special functions that can operate with *generic types*. This allows us to create a function template whose functionality can be adapted to more than one variable type or class without repeating the code for each type.

This is achieved through *template parameter*. A template parameter is a special kind of parameter that can be used to pass a type as parameter. These function templates can use these parameters as if they were regular types.

The format for declaring function templates with type parameters is :

```
template <class identifier> function_declaration;
template <typename identifier> function_declaration;
```

The only difference between both prototypes is the use of either the keyword class or the keyword typename. Its use is indistinct since both expressions have exactly the same meaning and behave exactly the same way.

For example, to create a template function that returns the greater one of two objects we could use :

```
template <class myType>
myType GetMax (myType a, myType b)
{
    return (a>b?a:b);
}
```

Here we have created a template function with myType as its template parameter. This template parameter represents a type that has not yet been specified, but that can be used in the template function as if it were a regular type. As you can see, the function template GetMax returns the greater of two parameters of this still-undefined type.

To use this function template we use the following format for the function call :

```
function_name <type> (parameters);
```

For example, to call GetMax to compare two integer values of type int we can write :

```
int x,y;
GetMax <int> (x,y);
```

When the compiler encounters this call to a template function, it uses the template to automatically generate a function replacing each appearance of myType by the type passed as the actual template parameter (int in this case) and then calls it. This process is automatically performed by the compiler and is invisible to the programmer.

Here is the entire example :

Example 8.1 **Function template**

```
#include <iostream.h>

template <class T>
T GetMax (T a, T b)
{
     T result;
     result = (a>b)? a : b;
     return (result);
}

int main ()
{
     int i = 5, j = 6, k;
     long l = 10, m = 5, n;

     k = GetMax<int>(i,j);
     n = GetMax<long>(l,m);

     cout << k << endl;
     cout << n << endl;

     return 0;
}
```

OUTPUT

```
6
10
```

In this case we have used T as the template parameter name instead of myType because it is shorter and in fact is a very common template parameter name. But you can use any identifier you like.

In the example above we used the function template GetMax() twice. The first time with arguments of type int and the second one with arguments of type long. The compiler has instantiated and then called each time the appropriate version of the function.

As you can see, the type T is used within the GetMax() template function even to declare new objects of that type :

```
T result;
```

Therefore, result will be an object of the same type as the parameters a and b when the function template is instantiated with a specific type.

In this specific case where the generic type T is used as a parameter for GetMax the compiler can find out automatically which data type has to instantiate without having to explicitly specify it within angle brackets (like we have done before specifying <int> and <long>). So we could have written instead :

```
int i,j;
GetMax (i,j);
```

Since both i and j are of type int, and the compiler can automatically find out that the template parameter can only be int. This implicit method produces exactly the same result :

Example 8.2 **Function template II**

```
#include <iostream.h>

template <class T>
T GetMax (T a, T b)
{
    return (a>b?a:b);
}

    int main ()
{
    int i = 5, j = 6, k;
    long l = 10, m = 5, n;

    k = GetMax(i,j);
    n = GetMax(l,m);

    cout << k << endl;
    cout << n << endl;

    return 0;
}
```

OUTPUT

```
6
10
```

Notice how in this case, we called our function template GetMax() without explicitly specifying the type between angle-brackets <>. The compiler automatically determines what type is needed on each call.

Because our template function includes only one template parameter (class T) and the function template itself accepts two parameters, both of this T type, we cannot call our function template with two objects of different types as arguments :

```
int i;
long l;
k = GetMax (i,l);
```

This would not be correct, since our GetMax function template expects two arguments of the same type, and in this call to it we use objects of two different types.

We can also define function templates that accept more than one type parameter, simply by specifying more template parameters. For example :

```
template <class T, class U>
T GetMin (T a, U b)
{
    return (a<b?a:b);
}
```

In this case, our function template GetMin() accepts two parameters of different types and returns an object of the same type as the first parameter (T) that is passed. For example, after that declaration we could call GetMin() with :

```
int i,j;
long l;
i = GetMin<int,long> (j,l);
```

or simply :

```
i = GetMin (j,l);
```

even though j and l have different types. But the compiler can determine the appropriate instantiation anyway.

8.3 CLASS TEMPLATES

We also have the possibility to write class templates, so that a class can have members that use template parameters as types. For example :

```
template <class T>
class pair
{
    T values [2];
public:
    pair (T first, T second)
    {
    values[0]=first; values[1]=second;
    }
};
```

The class that we have just defined serves to store two elements of any valid type. For example, if we wanted to declare an object of this class to store two integer values of type int with the values 115 and 36 we would write :

```
pair<int> myobject (115, 36);
```

this same class would also be used to create an object to store any other type :

```
pair<float> myfloats (3.0, 2.18);
```

The only member function in the previous class template has been defined inline within the class declaration itself. In case that we define a function member outside the declaration of the class template, we must always precede that definition with the template <...> prefix :

Example 8.3 **Class templates**

```
#include <iostream.h>

template <class T>
class pair
{
    T a, b;
    public:
    pair (T first, T second)
    {a=first; b=second;}
    T getmax ();
};

template <class T>
T pair<T>::getmax ()
{
    T retval;
    retval = a>b? a : b;
    return retval;
}

int main ()
{
    pair <int> myobject (100, 75);
    cout << myobject.getmax();
    return 0;
}
```

OUTPUT

```
100
```

Notice the syntax of the definition of member function getmax :

```
template <class T>
T pair<T>::getmax ()
```

Confused by so many T's? There are three T's in this declaration. The first one is the template parameter. The second T refers to the type returned by the function. And the third T (the one between angle brackets) is also a requirement. It specifies that this function's template parameter is also the class template parameter.

8.4 TEMPLATES AND INHERITANCE

A class template (or its instantiation) can serve as the base of a derived class :

```
template <class Type>
class SmartList : public List<Type>; // template base
class Primes : protected List<int>; // instantiated base
```

A template base class, such as List, should always be accompanied with its parameter list (or arguments if instantiated). The following is therefore invalid :

```
template <class Type>
class SmartList : public List; // illegal! <Type> missing
```

It would be equally incorrect to attempt to derive a non-template class from a (non-instantiated) template class :

```
class SmartList : public List<Type>; // illegal!template expected
```

It is, however, perfectly acceptable for a normal class to serve as the base of a derived template class :

```
class X;
template <class Type> class Y : X; // ok
```

8.5 TEMPLATES AND FRIENDS

When a function or class is declared as a friend of a class template, the friendship can take one of there forms, as illustrated by the following example.

Consider the Stack class template and a function template named Foo :

```
template <class T> void Foo (T&);
```

We wish to define a class named Sample and declare Foo and Stack as its friends. The following makes a *specific* instance of Foo and Stack friends of *all* instances of Sample :

```
template <class T>
class Sample { // one-to-many friendship
friend Foo<int>;
friend Stack<int>;
//...
};
```

Alternatively, we can make *each* instance of Foo and Stack a friend of its *corresponding* instance of Sample :

```
template <class T>
class Sample { // one-to-one friendship
friend Foo<T>;
```

```
friend Stack<T>;
//...
};
```

This means that, for example, Foo<int> and Stack<int> are friends of Sample<int>, but not Sample<double>.

The extreme case of making *all* instances of Foo and Stack friends of *all* instances of Sample is expressed as :

```
template <class T>
class Sample { // many-to-many friendship
template <class T> friend Foo;
template <class T> friend class Stack;
//...
};
```

The choice as to which form of friendship to use depends on the intentions of the programmer.

8.6 TEMPLATES AND STATIC MEMBERS

For static, inline, and extern functions, the respective keyword must appear after the template clause, and not before it :

```
template <class T>
inline T Max (T val1, T val2); // ok
inline template <class T> // illegal! inline misplaced
T Max (T val1, T val2);
```

Related Questions In Exams

Q.1. What do you understand by generic programming? What is a function template? *[May-2007]*

What is function templates? *[Dec-2007]*

What do you mean by Template? Write a program to demonstrate function template. *[May-2009]*

Ans. Refer Section 8.1 and 8.1.1

When we use the templates in our programs this is refer as a **Generic programming**.

Q.2. How overriding of template functions take place? *[Dec-2007]*

Ans. Let see an example, here we overriding an add function is shown :

```
#include<iostream.h>
template <class T>
class ABC
{
template <class T>
T add(T a, T b)
```

```
{
    return (a+b);
}
};
template <class T>
class XYZ: public ABC <T>
{
template <class T>
T add(T a, T b)
{
    return (a+b);
}
};

int main()
{
XYZ<int> X;
X.XYZ::add(10,20); //calling add function of XYZ class
X.ABC::add(20,20); //calling add function of ABC class
}
```

Q.3. What is class template & non-type parameters? *[May-2008]*

What do you mean by Template? Write a program to demonstrate class template. *[Dec-2006]*

Write a short note on Class template? *[May-2009]*

Ans. Refer Section 8.1 and 8.3.

Q.4. Write a function template for sorting arrays of various types. *[May-2007]*

Ans.

```
#include <iostream>
using namespace std;

template <class X> void bubble(X *items,int count)
{
    X t;

    for(int a=1; a<count; a++)
    for(int b=count-1; b>=a; b–)
    if(items[b-1] > items[b]) {
    t = items[b-1];
    items[b-1] = items[b];
    items[b] = t;
    }
}

int main()
```

```
{
    int iarray[7] = {7, 5, 4, 3, 9, 8, 6};
    double darray[5] = {4.3, 2.5, -0.9, 10.2, 3.0};

    cout << "Here is unsorted integer array: ";
    for(int i=0;  i<7; i++)
    cout << iarray[i] << ' ';
    cout << endl;

    bubble(iarray, 7);

    cout << "Here is sorted integer array: ";
    for(int i=0;  i<7; i++)
    cout << iarray[i] << ' ';
    cout << endl;

    cout << "Here is unsorted double array: ";
    for(int i=0;  i<5; i++)
    cout << darray[i] << ' ';
    cout << endl;

    bubble(darray, 5);

    cout << "Here is sorted double array: ";
    for(int i=0;  i<5; i++)
    cout << darray[i] << ' ';
    cout << endl;

    return 0;
}
```

Q.5. Differentiate between class template and template class. Write a generic program for the stack that takes as input int data and float data. *[Dec-2009]*

Ans. Class Template : Pre-defined class used to create template classes.

Template Class : User defined classes in generic using template class refer as template class.

```
#include <iostream>
using std::cout;
using std::endl;

template< typename T >
class Stack
{
public:
```

```
    template< typename T >
    Stack( int s ): size( s > 0 ? s : 10 ),top(-1 ), stackPtr( new T[ size ] )
    {
    }

   ~Stack()
    {
        delete [] stackPtr;
    }

    bool push( const T &pushValue )
    {
        if ( !isFull() )
    {
        stackPtr[ ++top ] = pushValue;
        return true;
    }

    return false;
    }
    bool pop( T &popValue )
    {
        if ( !isEmpty() )
        {
             popValue = stackPtr[ top– ];
             return true;
        }

        return false;
    }
        bool isEmpty() const
    {
        return top == -1;
    }

        bool isFull() const
    {
        return top == size - 1;
    }

private:
    int size;
    int top;
    T *stackPtr;
```

```
};

int main()
{
    Stack< double > doubleStack( 5 ); // size 5
    double doubleValue = 1.1;

    cout << "Pushing elements onto doubleStack\n";

    while ( doubleStack.push( doubleValue ) )
    {
        cout << doubleValue << ' ';
        doubleValue += 1.1;
    }

    while ( doubleStack.pop( doubleValue ) )
        cout << doubleValue << ' ';

    Stack< int > intStack; // default size 10
    int intValue = 1;
    cout << "\nPushing elements onto intStack\n";

    while ( intStack.push( intValue ) )
{
    cout << intValue << ' ';
    intValue++;
}

    while ( intStack.pop( intValue ) )
    cout << intValue << ' ';

    return 0;
}
```

8.7 EXCEPTION HANDLING

An **exception** is a runtime error. Proper handling of exceptions is an important programming issue. This is because exceptions can and do happen in practice and programs are generally expected to behave gracefully in face of such exceptions. Unless an exception is properly handled, it is likely to result in abnormal program termination and potential loss of work. For example, an undetected division by zero or de-referencing of an invalid pointer will almost certainly terminate the program abruptly.

Exceptions are basically runtime logical errors which are unexpected but we can predicate them.

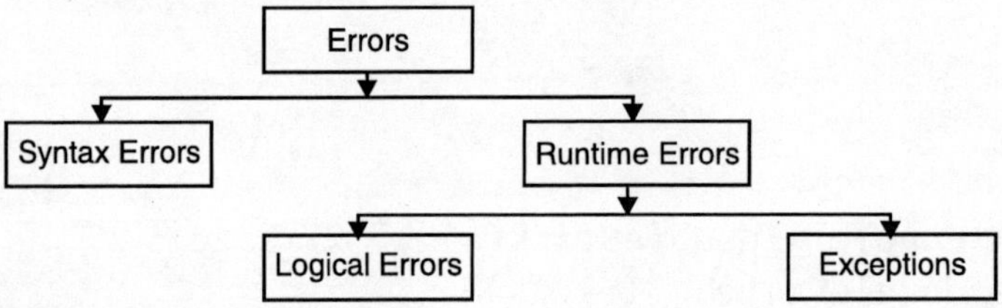

Errors are basically two types :

1. Syntax Errors
2. Runtime Errors

Syntax Errors : Syntax errors are those errors which user done by mistake in writing a syntax of statements of the source code of program. This type of errors can be caught at the time of compilation. And user can correct them after changing the syntax of code.

For example :

- Missing semi-colon,
- Missing of curly brackets,
- Undefined symbols

The above errors are the common errors, generally seen by all.

Runtime Errors : Runtime errors are those errors which can be shown after the execution of the program. This type of error can be shown when user gets unexpected answers or some errors which can be understand by the expert programmers.

The runtime errors are basically two types:

- Logical Errors.
- Exceptions.

Logical Errors : When user gets an unexpected answers it means some thing is wrong in the logic of program. That type of mistake is referred as logical errors.

For example :

- When you create a program to display number from 1 to10 but it shows from 0 to 9 or from 10 to 1.
- When you write one loop instead of two loops.
- When you done increment instead of decrement.

Exceptions : Exceptions are unexpected errors got at the time of execution. These type of errors can be cause at the time of wrong input or getting the wrong data type after calculation.

For example :

- Floating point error,
- Incompatible data type
- Data type mismatch
- Index out of bound.

In C++ exception can be handle by the concept of exception handling.

8.8 BASICS OF EXCEPTION HANDLING

Exception handling consists of three things: (i) the detecting of a run-time error, (ii) raising an exception in response to the error, and (ii) taking corrective action. The latter is called

recovery. Some exceptions can be fully recovered from so that execution can proceed unaffected. For example, an invalid argument value passed to a function may be handled by substituting a reasonable default value for it. Other exceptions can only be partially handled. For example, exhaustion of the heap memory can be handled by abandoning the current operation and returning to a state where other operations (such as saving the currently open files to avoid losing their contents) can be attempted.

try block

The try block is use to write those code of program which expected for the errors.

For example :

```
try
{
      :
      :
}
```

catch block

The catch block is used to receive the errors raise by the try block.

For example :

```
catch(int exception_code)
{
      :
      :
}
```

C++ provides a language facility for the uniform handling of exceptions. Under this scheme, a section of code whose execution may lead to runtime errors is labeled as a **try block**. Any fragment of code activated during the execution of a try block can raise an exception using a **throw clause**. All exceptions are typed (*i.e.*, each exception is denoted by an object of a specific type). A try block is followed by one or more **catch clauses**. Each catch clause is responsible for the handling of exceptions of a particular type.

When an exception is raised, its type is compared against the catch clauses following it. If a matching clause is found then its handler is executed. Otherwise, the exception is propagated up, to an immediately enclosing try block (if any). The process is repeated until either the exception is handled by a matching catch clause or it is handled by a default handler.

To catch exceptions we must place a portion of code under exception inspection. This is done by enclosing that portion of code in a *try block*. When an exceptional circumstance arises within that block, an exception is thrown that transfers the control to the exception handler. If no exception is thrown, the code continues normally and all handlers are ignored.

A exception is thrown by using the throw keyword from inside the try block. Exception handlers are declared with the keyword catch, which must be placed immediately after the try block :

Example 8.4 **Exceptions**

```
#include <iostream.h>
int main ()
```

```
{
        try
        {
        throw 20;
        }
        catch (int e)
        {
            cout << "An exception occurred. Exception Nr. " << e <<
            endl;
        }
        return 0;
}
```

OUTPUT

```
An exception occurred. Exception Nr. 20
```

The code under exception handling is enclosed in a try block. In this example this code simply throws an exception :

```
throw 20;
```

A throw expression accepts one parameter (in this case the integer value 20), which is passed as an argument to the exception handler.

The exception handler is declared with the catch keyword. As you can see, it follows immediately the closing brace of the try block. The catch format is similar to a regular function that always has at least one parameter. The type of this parameter is very important, since the type of the argument passed by the throw expression is checked against it, and only in the case they match, the exception is catch.

We can chain multiple handlers (catch expressions), each one with a different parameter type. Only the handler that matches its type with the argument specified in the throw statement is executed.

If we use an ellipsis (...) as the parameter of catch, that handler will catch any exception no matter what the type of the throw exception is. This can be used as a default handler that catches all exceptions not catch by other handlers if it is specified at last :

```
try
{
    // code here
}
    catch (int param) { cout << "int exception"; }
    catch (char param) { cout << "char exception"; }
    catch (...) { cout << "default exception"; }
```

In this case the last handler would catch any exception thrown with any parameter that is neither an int nor a char.

After an exception has been handled the program execution resumes after the try-catch block, not after the throw statement!.

It is also possible to nest try-catch blocks within more external try blocks. In these cases, we have the possibility that an internal catch block forwards the exception to its external level. This is done with the expression throw; with no arguments. For example :

```
try {
    try {
        // code here
            }
        catch (int n)
        {
            throw;
        }
    }
        catch (...)
        {
            cout << "Exception occurred";
        }
```

8.9 RE-THROWING AN EXCEPTION

If a catch block cannot handle the particular exception it has caught, you can rethrow the exception. The rethrow expression (throw without *assignment_expression*) causes the originally thrown object to be rethrown.

Because the exception has already been caught at the scope in which the rethrow expression occurs, it is rethrown out to the next dynamically enclosing try block. Therefore, it cannot be handled by catch blocks at the scope in which the rethrow expression occurred. Any catch blocks for the dynamically enclosing try block have an opportunity to catch the exception.

The following example demonstrates rethrowing an exception :

```
#include <iostream>
using namespace std;

struct E {
    const char* message;
    E() : message("Class E") { }
};

struct E1 : E {
    const char* message;
    E1() : message("Class E1") { }
};

struct E2 : E {
    const char* message;
    E2() : message("Class E2") { }
};

void f() {
```

```
    try {
        cout << "In try block of f()" << endl;
        cout << "Throwing exception of type E1" << endl;
        E1 myException;
        throw myException;
    }
    catch (E2& e) {
        cout << "In handler of f(), catch (E2& e)" << endl;
        cout << "Exception: " << e.message << endl;
        throw;
    }
    catch (E1& e) {
        cout << "In handler of f(), catch (E1& e)" << endl;
        cout << "Exception: " << e.message << endl;
        throw;
    }
    catch (E& e) {
        cout << "In handler of f(), catch (E& e)" << endl;
        cout << "Exception: " << e.message << endl;
        throw;
    }
}

int main() {
    try {
        cout << "In try block of main()" << endl;
        f();
    }
    catch (E2& e) {
        cout << "In handler of main(), catch (E2& e)" << endl;
        cout << "Exception: " << e.message << endl;
    }
    catch (...) {
        cout << "In handler of main(), catch (...)" << endl;
    }
}
```

The following is the output of the above example :

```
In try block of main()
In try block of f()
Throwing exception of type E1
In handler of f(), catch (E1& e)
Exception: Class E1
In handler of main(), catch (...)
```

The try block in the main() function calls function f(). The try block in function f() throws an object of type E1 named `myException`. The handler `catch (E1 &e)` catches `myException`. The handler then rethrows `myException` with the statement `throw` to the next dynamically enclosing try block: the try block in the `main()` function. The handler `catch(...)` catches `myException`.

8.10 EXCEPTION SPECIFICATIONS

When declaring a function we can limit the exception type it might directly or indirectly throw by appending a throw suffix to the function declaration :

```
float myfunction (char param) throw (int);
```

This declares a function called myfunction which takes one agument of type char and returns an element of type float. The only exception that this function might throw is an exception of type int. If it throws an exception with a different type, either directly or indirectly, it cannot be catch by a regular int-type handler.

If this throw specifier is left empty with no type, this means the function is not allowed to throw exceptions. Functions with no throw specifier (regular functions) are allowed to throw exceptions with any type :

```
int myfunction (int param) throw(); // no exceptions allowed
int myfunction (int param);         // all exceptions allowed
```

8.11 PROCESSING UNEXPECTED EXCEPTIONS

When a function with an exception specification throws an exception that is not listed in its exception specification, the C++ runtime does the following:

1. The unexpected() function is called.
2. The unexpected() function calls the function pointed to by unexpected_handler. By default, unexpected_handler points to the function terminate().

You can replace the default value of unexpected_handler with the function set_unexpected().

Although unexpected() cannot return, it may throw (or rethrow) an exception. Suppose the exception specification of a function f() has been violated. If unexpected() throws an exception allowed by the exception specification of f(), then the C++ run time will search for another handler at the call of f(). The following example demonstrates this:

Example 8.5 **Unexpected Exception**

```
#include <iostream>
using namespace std;

struct E {
    const char* message;
    E(const char* arg) : message(arg) { }
};

void my_unexpected() {
    cout << "Call to my_unexpected" << endl;
```

```
    throw E("Exception thrown from my_unexpected");
}

void f() throw(E) {
    cout << "In function f(), throw const char* object" << endl;
    throw("Exception, type const char*, thrown from f()");
}

int main() {
    set_unexpected(my_unexpected);
    try {
        f();
    }
    catch (E& e) {
        cout << "Exception in main(): " << e.message << endl;
    }
}
```

OUTPUT

```
In function f(), throw const char* object
Call to my_unexpected
Exception in main(): Exception thrown from my_unexpected
```

The `main()` function's try block calls function `f()`. Function `f()` throws an object of type `const char*`. However the exception specification of `f()` allows only objects of type `E` to be thrown. The function `unexpected()` is called. The function `unexpected()` calls `my_unexpected()`. The function `my_unexpected()` throws an object of type `E`. Since `unexpected()` throws an object allowed by the exception specification of `f()`, the handler in the `main()` function may handle the exception.

If `unexpected()` did not throw (or rethrow) an object allowed by the exception specification of `f()`, then the C++ run time does one of two things:

- If the exception specification of `f()` included the class `std::bad_exception`, `unexpected()` will throw an object of type `std::bad_exception`, and the C++ run time will search for another handler at the call of `f()`.
- If the exception specification of `f()` did not include the class `std::bad_exception`, the function `terminate()` is called.

8.12 STACK UNWINDING

When an exception is thrown and control passes from a try block to a handler, the C++ runtime calls destructors for all automatic objects constructed since the beginning of the try block. This process is called *stack unwinding.* The automatic objects are destroyed in reverse order of their construction. (Automatic objects are local objects that have been declared auto or register, or not declared static or extern. An automatic object x is deleted whenever the program exits the block in which x is declared.)

If an exception is thrown during construction of an object consisting of subobjects or array elements, destructors are only called for those subobjects or array elements successfully constructed before the exception was thrown. A destructor for a local static object will only be called if the object was successfully constructed.

If during stack unwinding a destructor throws an exception and that exception is not handled, the terminate() function is called. The following example demonstrates this:

Example 8.6 **Stack Unwinding**

```
#include <iostream>
using namespace std;

struct E {
   const char* message;
   E(const char* arg) : message(arg) { }
};

void my_terminate() {
   cout << "Call to my_terminate" << endl;
};

struct A {
   A() { cout << "In constructor of A" << endl; }
   ~A() {
      cout << "In destructor of A" << endl;
      throw E("Exception thrown in ~A()");
   }
};

struct B {
   B() { cout << "In constructor of B" << endl; }
   ~B() { cout << "In destructor of B" << endl; }
};

int main() {
   set_terminate(my_terminate);

   try {
      cout << "In try block" << endl;
      A a;
      B b;
      throw("Exception thrown in try block of main()");
   }
   catch (const char* e) {
      cout << "Exception: " << e << endl;
   }
   catch (...) {
      cout << "Some exception caught in main()" << endl;
   }
```

```
    cout << "Resume execution of main()" << endl;
}
```

OUTPUT

```
In try block
In constructor of A
In constructor of B
In destructor of B
In destructor of A
Call to my_terminate
```

In the try block, two automatic objects are created: `a` and `b`. The try block throws an exception of type `const char*`. The handler `catch (const char* e)` catches this exception. The C++ run time unwinds the stack, calling the destructors for `a` and `b` in reverse order of their construction. The destructor for `a` throws an exception. Since there is no handler in the program that can handle this exception, the C++ run time calls `terminate()`. (The function `terminate()` calls the function specified as the argument to `set_terminate()`. In this example, `terminate()` has been specified to call `my_terminate()`.)

8.13 EXCEPTION AND INHERITANCE

Since it's possible to throw classes as exceptions, and classes can be derived from other classes, we need to consider what happens when we use inherited classes as exceptions. As it turns out, exception handlers will not only match classes of a specific type, they'll also match classes derived from that specific type as well! Consider the following example:

Example 8.7 **Exception with Inheritance**

```
class Base
{
public:
      Base() {}
};

class Derived: public Base
{
public:
      Derived() {}
};
int main()
{
      try
      {
            throw Derived();
      }
```

```
        catch (Base &cBase)
        {
            cerr << "caught Base";
        }
        catch (Derived &cDerived)
        {
            cerr << "caught Derived";
        }

        return 0;
    }
```

OUTPUT:

```
caught Base
```

First, as mentioned above, derived classes will be caught by handlers for the base type. Because Derived is derived from Base, Derived is-a Base (they have an is-a relationship). Second, when C++ is attempting to find a handler for a raised exception, it does so sequentially. Consequently, the first thing C++ does is check whether the exception handler for Base matches the Derived exception. Because Derived is-a Base, the answer is yes, and it executes the catch block for type Base! The catch block for Derived is never even tested in this case.

In order to make this example work as expected, we need to flip the order of the catch blocks:

Example 8.8 Exception with Inheritance

```
class Base
{
public:
        Base() {}
};

class Derived: public Base
{
public:
        Derived() {}
};
int main()
{
        try
        {
            throw Derived();
        }
        catch (Derived &cDerived)
        {
            cerr << "caught Derived";
        }
        catch (Base &cBase)
```

```
        {
                cerr << "caught Base";
        }
        return 0;
}
```

This way, the Derived handler will get first shot at catching objects of type Derived (before the handler for Base can). Objects of type Base will not match the Derived handler (Derived is-a Base, but Base is not a Derived), and thus will "fall through" to the Base handler.

Rule: Handlers for derived exception classes should be listed before those for base classes.

The ability to use a handler to catch exceptions of derived types using a handler for the base class turns out to be exceedingly useful.

Let's take a look at this using std::exception. There are many classes derived from std::exception, such as std::bad_alloc, std::bad_cast, std::runtime_error, and others. When the standard library has an error, it can throw a derived exception correlating to the appropriate specific problem it has encountered.

Most of the time, we probably won't care whether the problem was a bad allocation, a bad cast, or something else. We just care that we got an exception from the standard library. In this case, we just set up an exception handler to catch std::exception, and we'll end up catching std::exception and all of the derived exceptions together in one place. Easy!

```
try
{
   // code using standard library goes here
}
// This handler will catch std::exception and all the derived
exceptions too
     catch (std::exception &cException)
{
     cerr << "Standard exception: " << cException.what() << endl;
}
```

However, sometimes we'll want to handle a specific type of exception differently. In this case, we can add a handler for that specific type, and let all the others "fall through" to the base handler. Consider:

```
   try
   {
          // code using standard library goes here
   }
   // This handler will catch std::bad_alloc (and any exceptions derived
from it) here
   catch (std::bad_alloc &cException)
   {
       cerr << "You ran out of memory!" << endl;
   }
```

```
    // This handler will catch std::exception (and any exception derived
from it) that fall
    // through here
    catch (std::exception &cException)
    {
        cerr << "Standard exception: " << cException.what() << endl;
    }
```

In this example, exceptions of type std::bad_alloc will be caught by the first handler and handled there. Exceptions of type std::exception and all of the other derived classes will be caught by the second handler.

Such inheritance hierarchies allow us to use specific handlers to target specific derived exception classes, or to use base class handlers to catch the whole hierarchy of exceptions. This allows us a fine degree of control over what kind of exceptions we want to handle while ensuring we don't have to do too much work to catch "everything else" in a hierarchy.

Related Questions In Exams (Most Important)

Q.1. What is an exception? How is an exception handled in C++? *[Dec-2006]*

What is an exception? Give some reasons that causes exceptions? *[May-2009]*

What is exception handling? *[Dec-2009]*

Write a short note on exception handling. *[Dec-2009]*

Ans. Refer Section 8.7 and 8.8.

Q.2.What are exception handlers? *[Dec-2007]*

What are exception handlers? *[May-2008]*

What is exception handler? What are the steps involved in exception handling? *[Dec-2008]*

Ans. Suppose we have an application that works with objects of a certain class. If during the course of execution of a member function of this class an error occurs, then this member function informs the application that an error has occurred. This process of informing is called throwing an exception.

In the application we have to create a separate section of code to tackle the error. This section of code is called an exception handler or a catch block. Any code in the application that uses objects of the class is enclosed in a try block. Errors generated in the try block are caught in the catch block. Code that doesn't interact with the class need not be in a try blocks. It is not a working program, but it clearly shows how and where the various elements of the exception mechanism are placed.

Q.3. Write a program to demonstrate the concept of re-throwing an exception. *[Dec-2006]*

How re-throwing an exception is being done? *[May-2008]*

Ans. Refer Section 8.9.

Q.4. Explain what happens when an exception is thrown in C++? *[Dec-2009]*

Ans. Always exception thrown by the try block and after thrown exception it caught by the associated catch block.

Q.5. What happens if an exception is not caught? *[Dec-2009]*

Ans. Then program will terminate abnormally.

Q.6. What happens if an exception is thrown from an object's constructor? *[Dec-2009]*

Ans. It can be handle by the catch block. C++ provides specific error code for fail to object creation.

Q.7. What happens if an exception is thrown from an object's destructor?
[Dec-2009]

Ans. It can be handle by the catch block. C++ provides specific error code for fail to release the object.

Q.8. What are the costs and benefits of using exceptions? *[Dec-2009]*

Ans. Benefits :

- Provide robust code.
- Save from abnormal termination
- Provide readable code.
- Provides centralised error handling

Costs :

- Error recovery is more complex.
- Error cause will terminate the program some message which is not expected by the user.

Q.9. When would you choose to return an error code rather than throwing an exception? *[Dec-2009]*

Ans.
- When error is easy to resolve.
- When error is predefine in C++ compiler and can be use the message associated with code.

Summary

The template is one of C++'s most sophisticated and high-powered features. Although not part of the original specification for C++, it was added several years ago and is supported by all modern C++ compilers. Using templates, it is possible to create generic functions and classes. In a generic function or class, the type of data upon which the function or class operates is specified as a parameter. Thus, you use one function or class with seveal different types of data without having to explicitly recode specific versions for each data type. We have covered both in this chapter.

This chapter is also discussed the exception handling subsystem. Exception handling allows you to manage run-time errors in an orderly fashion. Using exception handling, your program can automatically invoke an error-handling routine when an error occurs. The principal advantage of exception handling is that it automates much of the error-handling code that previously had to be coded "by hand" in any large program.

Exercise

Q. 8.1. What are the advantages of templates?

Q. 8.2. Explain function template.

Q. 8.3. What is the syntax of class template? Explain with example.

Q. 8.4. What do you mean by Exceptions? How can we handle exceptions?

Q. 8.5. Explain the 'try' and 'catch' blocks.

❑❑❑

C++ PROGRAMMING LAB.

C++ PROGRAMMING LAB.

Q.1. Raising a number n to a power p is the same as multiplying n by itself p times. Write a function called power() that takes a double value for n and an int value for p, and returns the result as double value. Use a default argument of 2 for p, so that if this argument is omitted, the number will be squared. Write a main() function that gets values from the user to test this function.

```
#include<iostream.h>
#include<conio.h>
double power(double n,int p=2)
{
    double a=1.0;
    for(int i=1;i<=p;i++)
        a=a*n;
    return a;
}
void main()
{
    clrscr();
    double number, ans;
    int pow;
    cout<<"\n Enter number for calculation:"<<endl;
    cin>>number;
    cout<<"\n Enter power for "<<number<<" :";
    cin>>pow;
    ans=power(number,pow);
    cout<<"\n Answer for "<<number<<"^"<<pow<<" : "<<ans;
    ans=power(number);
    cout<<"\n Answer for "<<number<<" without sending power "<<"
    :"<<ans;
}
```

Q2. A point on the two-dimensional plane can be represented by two numbers: an X coordinate and a Y coordinate. For example, (4,5) represents a point 4 units to the right of the origin along the X axis and 5 units up the Y axis. The sum of two points can be defined as a new point whose X coordinate is the sum of the X coordinates of the points and whose Y coordinate is the sum of their Y coordinates.

Write a program that uses a structure called point to model a point. Define three points, and have the user input values to two of them. Than set the third point equal to the sum of the other two, and display the value of the new point. Interaction with the program might look like this :

Enter coordinates for P1: 3 4

Enter coordinates for P2: 5 7

Coordinates of P1 + P2 are : 8, 11

```
#include<iostream.h>
#include<conio.h>
typedef struct point
{
    int x,y;
public:
    void getvalues()
    {
        cin>>x>>y;
    }
    void show()
    {
        cout<<x<<","<<y;
    }
}point;
point add(point first, point second)
{
    point third;
    third.x=first.x + second.x;
    third.y=first.y + second.y;
    return third;
}

void main()
{
    clrscr();
    point p1,p2,p3;
    cout<<"\nEnter coordinates for P1"<<endl;
```

```
        p1.getvalues();
        cout<<"\nEnter coordinates for P2"<<endl;
        p2.getvalues();
        p3=add(p1,p2);
        cout<<"co-ordinates of P1+P2 are ";
        p3.show();
    }
```

Q. 3. Create the equivalent of a four function calculator. The program should request the user to enter a number, an operator, and another number. It should then carry out the specified arithmetical operation: adding, subtracting, multiplying, or dividing the two numbers. (It should use a switch statement to select the operation). Finally it should display the result. When it finishes the calculation, the program should ask if the user wants to do another calculation. The response can be 'Y' or 'N'. Some sample interaction with the program might look like this.

Enter first number, operator, second number: 10/ 3 Answer = 3.333333
Do another (Y/ N)? Y
Enter first number, operator, second number 12 + 100
Answer = 112
Do another (Y/ N) ? N

```
    #include<iostream.h>
    #include<conio.h>
    int main()
    {
        double op1,op2,ans;
        char opr, choice;
        clrscr();
        do
        {
        cout<<"\nEnter first number, operator, second number :";
        cin>>op1>>opr>>op2;
        switch(opr)
        {
            case '+':ans =op1+op2; break;
            case '-':ans =op1-op2; break;
            case '*':ans =op1*op2; break;
            case '/':ans =op1/op2; break;
            default: cout<<"\nInvalid operator plz check!!!"<<endl;
                }
        cout<<"\nAnswer = "<<ans;
```

```
cout<<"\nDo another (Y/N)?: ";
cin>>choice;
}while(choice=='Y' || choice=='y');

return 0;
}
```

Q.4. A phone number, such as (212) 767-8900, can be thought of as having three parts: the area code (212), the exchange (767) and the number (8900). Write a program that uses a structure to store these three parts of a phone number separately. Call the structure phone.

Create two structure variables of type phone. Initialize one, and have the user input a number

for the other one. Then display both numbers. The interchange might look like this:

Enter your area code, exchange, and number: 415 555 1212
My number is (212) 767-8900
Your number is (415) 555-1212

```
#include<iostream.h>
#include<conio.h>
struct Phone
{
int code,exchange,number;
};
int main()
{
struct Phone p1={212,767,8900};
struct Phone p2;
clrscr();
cout<<"\n Enter your area code, exchange and number: ";
cin>>p2.code>>p2.exchange>>p2.number;
cout<<"\n My number is ("<<p1.code<<")"<<p1.exchange<<"-"<<p1.number;
cout<<"\n Your number is ("<<p2.code<<")"<<p2.exchange<<"-"<<p2.number;
return 0;
}
```

Q.5. Create two classes DM and DB which store the value of distances. DM stores distances in metres and centimeters and DB in feet and inches. Write a program that can read values for the class objects and add one object of DM with another object of DB.

Use a friend function to carry out the addition operation. The object that stores the results maybe a DM object or DB object, depending on the units in which the results are required.

The display should be in the format of feet and inches or metres and cenitmetres depending on the object on display.

```
#include<iostream.h>
#include<conio.h>
class DB;
class DM
{
double m, cm;
//m:meter, cm:centimeter
public:
DM()
{
m=0.0;
cm=0.0;
}
void get_distance()
{
cout<<"\nEnter distance in meter and centimeters ";
cout<<"\n Meters ="; cin>>m;
cout<<"\n Centimeters ="; cin>>cm;
}
void show_distance()
{
cout<<"\n"<<m<<" meters and "<<cm<<" centimeters"<<endl;
}
friend void add_distance(DM D1, DB D2);
};
class DB
{
double f,in;
//f:feet, in:inches
public:
DB()
{
f=0.0;
in=0.0;
```

```
}
void get_distance()
{
cout<<"\nEnter distance in Feet and Inches ";
cout<<"\n Feet ="; cin>>f;
cout<<"\n Inches ="; cin>>in;
}
void show_distance()
{
cout<<"\n"<<f<<" Feet and "<<in<<" inches"<<endl;
}
friend void add_distance(DM D1, DB D2);
};
void add_distance(DM D1, DB D2)
{
    int choice;
    cout<<"\nEnter choice for calculation:";
    cout<<"\n Press 1 for distance in meters";
    cout<<"\n Press 2 for distance in meters";
    cin>>choice;
    if(choice==1)
    {
        double ftm,itoc;
        ftm=D2.f/3.3; // a meter = 3.3 feets
        itoc=D2.in*2.5; // an inches = 2.5 centimeters
        D1.m=D1.m+ftm;
        D1.cm=D1.cm+itoc;
        cout<<"D1.m:"<<D1.m<<endl;
        cout<<"D1.cm:"<<D1.cm<<endl;
    if(D1.cm>100.0) // 1 meter = 100 centimeters
    {
        D1.m+=D1.cm/100.0;
        D1.cm=D1.cm-100.0;
    }
    D1.show_distance();
}
if(choice==2)
{
    double mtf,ctoi;
```

```
        mtf=D1.m*3.3;                    // a meter = 3.3 feets
        ctoi=D1.cm/2.5;                  // an inches = 2.5 centimeters
        D2.f=D2.f+mtf;
        D2.in=D2.in+ctoi;
        if(D2.in>12.0)                   //1 feet = 12 inches
        {
            D2.f+=D2.in/100.0;
            D2.in=D2.in-100.0;
        }
        D2.show_distance();
    }
    }
    int main()
    {
        clrscr();
        DM d1;
        DB d2;
        d1.get_distance();
        d2.get_distance();
        cout<<"\n Distance after addition =";
        add_distance(d1,d2);
        return 0;
    }
```

Q.6. Create a class rational which represents a numerical value by two double values NUMERATOR & DENOMINATOR. Include the following public member Functions:

- **constructor with no arguments (default).**
- **constructor with two arguments.**
- **void reduce( ) that reduces the rational number by eliminating the highest common factor between the numerator and denominator.**
- **Overload + operator to add two rational number.**
- **Overload >> operator to enable input through cin.**
- **Overload << operator to enable output through cout.**
- **Write a main ( ) to test all the functions in the class.**

```
#include<iostream.h>
#include<conio.h>
#include<process.h>
class rational
{
    private:
```

```
        int num;
        int den;
        int lcm(int a,int b);
        int hcf(int a,int b);
    public:
    rational()
    {
        num=0;
        den=1;
    }
    rational(int n,int d=1)
    {
        if(d==0)
        {
            cout<<"denominator can't be zero";
            getch();
            exit(1);
        }
        else
        {
            num=n;
            den=d;
        }
    }
    rational reduce()
    {
        rational temp;
        int h=hcf(num,den);
        temp.num=num/h;
        temp.den=den/h;
        return temp;
    }
    rational operator +(rational r)
    {
        int l,t1,t2,x1,x2,den1,den2,p;
        rational ans;
        if(den<0)
            den1=-den;
        else
```

```
        den1=den;
    if(r.den<0)
        den2=-r.den;
    else
        den2=r.den;
    p=lcm(den1,den2);
    if((den<0&&r.den>0)||(den>0&&r.den<0))
        l=-p;
    else
        l=p;
    ans.den=l;
    t1= l/den;
    t2=l/r.den;
    x1=num*t1;
    x2=r.num*t2;
    ans.num=x1+x2;
    return ans;
}
fri end istream& operator >> (istream& s,rational& r);
friend ostream& operator << (ostream& s,rational& r);
};
int rational::lcm(int a,int b)
{
    int i,lcm=1;
    while(!(a==1&&b==1))
    {
        i=2;
        while(!(a%i==0||b%i==0)&&i<(a>b?a:b))
        i++;
        lcm*=i;
        if(a%i==0)
        a/=i;
        if(b%i==0)
        b/=i;
    }
    return lcm;
}
int rational::hcf(int a,int b)
{
    int r=a%b;
```

```
    while(r)
    {
        a=b;
        b=r;
        r=a%b;
    }
    return b;
}
istream& operator >>(istream& s,rational& r)
{
    int a,b;
    char c;
    s>>a>>c>>b;
    if(c!='/')
    {
        cout<<"use of invalid notation";
        getch();
        exit(0);
    }
    if(b==0)
    {
        cout<<"denominator can't be zero.";
        getch();
        exit(1);
    }
    r.num=a;
    r.den=b;
    return s;
}
ostream& operator <<(ostream& s,rational& r)
{
    if(r.den==1)
        s<<r.num;
    else
    {
        if(r.den==-1)
            s<<-r.num;
        else
            s<<r.num<<'/'<<r.den;
```

```
    }
    return s;
}
int main()
{
    clrscr();
    rational r1,r2,r3;
    cout<<"enter r1:";
    cin>>r1;
    cout<<"enter r2:";
    cin>>r2;
    r3=r1+r2;
    cout<<r1<<" + "<<r2<<" = "<<r3<<" = "<<r3.reduce();
    getch();
    return 0;
}
```

Q.7. Consider the following class definition

```
class father {
protected : int age;
public:
father (int x) {age = x;}
virtual void iam ( )
{ cout < < "I AM THE FATHER, my age is : "<< age<< endl:} };
```

Derive the two classes son and daughter from the above class and for each, define iam() to write our similar but appropriate messages. You should also define suitable constructors for these classes.

Now, write a main() that creates objects of the three classes and then calls iam() for them. Declare pointer to father. Successively, assign addresses of objects of the two derived classes to this pointer and in each case, call iam() through the pointer to demonstrate polymorphism in action.

```
#include<iostream.h>
#include<conio.h>
class father
{
protected:
    int age;
public:
    father(){age=0;}
    father(int x){age=x;}
```

```
        virtual void iam()
        {
        cout<<"I AM THE FATHER, My age is : "<<age<<endl;
        }
    };
    class son:public father
    {
    public:
        son(int x):father(x){}
        void iam()
        {
        cout<<"I AM THE SON, My age is : "<<age<<endl;
        }
    };
    class daughter:public father
    {
    public:
        daughter(int x):father(x){}
        void iam()
        {
        cout<<"I AM THE DAUGHTER, My age is : "<<age<<endl;
        }
    };

    int main()
    {
        clrscr();
        father f(60);
        son s(25);
        daughter d(20);
        f.iam();
        s.iam();
        d.iam();

        father *f1;
        son s1(25);
        daughter d1(20);
        f1=&s1;
        f1->iam();
        f1=&d1;
```

```
        f1->iam();

        return 0;
    }
```

Q.8. Write a program that creates a binary file by reading the data for the students from the terminal.

The data of each student consist of roll no., name (a string of 30 or lesser no. of characters) and marks.

```
    #include<fstream.h>
    #include<conio.h>
    class student
    {
    int rollno;
    char name[30];
    int marks;
    public:
    void getdata()
        {
        cout<<"\n Enter the Roll no: ";
        cin>>rollno;
        cout<<"\n Enter the Name: ";
        cin>>name;
        cout<<"\n Enter the Marks: ";
        cin>>marks;
        }
    void putdata()
        {
        cout<<"\n Roll no: "<<rollno;
        cout<<"\n Name: "<<name;
        cout<<"\n Marks: "<<marks;
        }
    };

    int main()
    {
        student S;
        fstream fin;
        fin.open("student.data",ios::out);
```

```
    for(int i=0;i<5;i++)
{
    S.getdata();
    fin.write((char*) & S, sizeof S);
}
fin.close();
return 0;
}
```

Q.9. A hospital wants to create a database regarding its indoor patients. The information to store include

(*a*) Name of the patient

(*b*) Date of admission

(*c*) Disease

(*d*) Date of discharge

Create a structure to store the date (year, month and date as its members). Create a base class to store the above information. The member function should include functions to enter information and display a list of all the patients in the database. Create a derived class to store the age of the patients.

List the information about all the patient to store the age of the patients.

List the information about all the pediatric patients (less than twelve years in age).

```
#include<iostream.h>
#include<conio.h>
struct Date
{ int dd,mm,yy;};
class patient
{
char name[30],disease;
Date adm_dt,dis_dt;
public:
void getdata()
    {
    cout<<"\n Enter the Name of Patient: ";
    cin>>name;
    cout<<"\n Enter the date of admission:(dd mm yyyy) ";
    cin>>adm_dt.dd>>adm_dt.mm>>adm_dt.yy;
    cout<<"\n Enter the Disease: ";
    cin>>disease;
    cout<<"\n Enter the date of discharge:(dd mm yyyy) ";
```

```
        cin>>dis_dt.dd>>dis_dt.mm>>dis_dt.yy;
        }
void putdata()
        {
        cout<<"\n Name: "<<name;
        cout<<"\n Diseases: "<<disease;
        cout<<"\nAdmission Date:"<<adm_dt.dd<<"-"<<adm_dt.mm<<"-
        <<adm_dt.yy;
        cout<<"\nDate of Discharge:"<<dis_dt.dd<<"-"<<dis_dt.mm<<"-
        "<<dis_dt.yy;
        }
    };
    class Person:public patient
    {
    int age;
    public:
    void getdata()
    {
        patient::getdata();
        cout<<"\nEnter the Age:";
        cin>>age;
    }
    void putdata()
    {
        patient::putdata();
        cout<<"\nAge:"<<age;
    }
    void pediatric_patient()
    {
        if(age<12)
        patient::putdata();
    }
    };
    int main()
    {
        Person P[10];
        clrscr();
        for(int i=0;i<2;i++)
        {
            P[i].getdata();
```

```
    }
cout<<"\n List of all Patients:"<<endl;
    for(i=0;i<5;i++)
    {
        P[i].putdata();
    }
cout<<"\n List of all pediatric patients:"<<endl;
    for(i=0;i<5;i++)
    {
        P[i].pediatric_patient();
    }
return 0;
}
```

Q.10. Make a class *Employee* with a name and salary. Make a class *Manager* inherit from *Employee*. Add an instance variable, named department, of type string. Supply a method to toString that prints the manager's name, department and salary. Make a class *Executive* inherit from Manager. Supply a method toString that prints the string "Executive" followed by the information stored in the *Manager* superclass object. Supply a test program that tests these classes and methods.

```
#include<iostream.h>
#include<conio.h>
class Employee
{
protected:
    char name[10];
    int salary;
};
class Manager:public Employee
{
    char dept[10];
public:
    void getdata()
    {
        cout<<"\n Enter Name:"; cin>>name;
        cout<<"\nEnter Department:"; cin>>dept;
        cout<<"\nEnter Salary:"; cin>>salary;
    }
    void toString()
    {
```

```
        cout<<"\n Name:"<<name;
        cout<<"\nDepartment:"<<dept;
        cout<<"\nSalary:"<<salary;
    }
};
class Executive:public Manager
{
 public:
    void toString()
    {
        cout<<"\n Executive ";
        Manager::toString();
    }
};

int main()
{
    clrscr();
    //using object of Manager Class
    Manager M;
    M.getdata();
    M.toString();

    //using object of Executive class
    Executive E;
    E.getdata();
    E.toString();
    getch();
    return 0;
}
```

Q.11. Imagine a toll booth with a class called toll Booth. The two data items are a type unsigned int to hold the total number of cars, and a type double to hold the total amount of money collected. A constructor initialises both these to 0. A member function called payingCar ( ) increments the car total and adds 0.50 to the cash total. Another function, called nopayCar ( ), increments the car total but adds nothing to the cash total. Finally, a member function called displays the two totals.

Include a program to test this class. This program should allow the user to push one key to count a paying car, and another to count a nonpaying car. Pushing the ESC key should cause the program to print out the total cars and total cash and then exit.

```
#include<iostream.h>
#include<conio.h>
class toll_Booth
{
    unsigned int tot_cars;
    double tot_amt;
    public:
       toll_Booth(){ tot_cars=0; tot_amt=0.00;}
    void payingCar()
       {
       tot_cars++;
           tot_amt+=0.50;
       }
       void nopayCar()
       {
       tot_cars++;
       }
       void display()
       {
           cout<<"\n Total Cars = "<<tot_cars;
           cout<<"\n Total Cash = "<<tot_amt;
       }

};
int main()
{
    clrscr();
    char choice;
    toll_Booth T;
    //Menu for user choice
    clrscr();
    do
    {
       cout<<"\n\n\n Push 1 to count Paying Cars.";
       cout<<"\nPush 2 to count Non Paying Cars.";
       cout<<"\nPush Esc to Exit & Display."<<endl;
       choice=getche();
       if(choice=='1') T.payingCar();
       if(choice=='2') T.nopayCar();
```

```
    }while((int)choice!=27); //27 is ascii code of ESC key
    T.display();
    getch();
    return 0;
}
```

Q.12. Write a function called reversit ( ) that reverses a string (an array of char). Use a for loop that swaps the first and last characters, then the second and next to last characters and so on. The string should be passed to reversit ( ) as an argument.

Write a program to exercise reversit ( ). The program should get a string from the user, call reversit ( ), and print out the result. Use an input method that allows embedded blanks. Test the program with Napoleon's famous phrase, "Able was I ere I saw Elba".

```
#include<iostream.h>
#include<conio.h>
#include<string.h>
void reversit(char* s)
{
    int len=strlen(s);
    char temp[40];
    len–;
    for(int i=0;i<=len;i++)
    {
        temp[i]=s[len-i];
    }
    temp[i]='\0';
    cout<<"\n Reverse String = "<<temp;
}
int main()
{
    clrscr();
    char str[40];
    cout<<"\n Enter the string : ";
    getline(cin, str);
    reversit(str);
    return 0;
}
```

Q.13. Create some objects of the string class, and put them in a Deque -some at the head of the Deque and some at the tail. Display the contents of the Deque using the

forEach ( ) function and a user written display function. Then search the Deque for a particular string, using the firstThat() function and display any strings that match. Finally remove all the items from the Deque using the getLeft() function and display each item. Notice the order in which the items are displayed: Using getLeft(), those inserted on the left (head) of the Deque are removed in "last in first out" order while those put on the right side are removed in "first in first out" order. The opposite would be true if getRight() were used.

```
#include<iostream.h>
#include<conio.h>
#include<string.h>
class String
{
    char str[20];
public:
    String(){ strcpy(str," ");}
    String(char* s){ strcpy(str,s);}
    void show(){cout<<str<<endl;}
    int same(char* s){ return(strcmp(s,str));}
};
class Deque
{
    String arr[20];
    int top,bottom;
    public:
    Deque(){ top=0;bottom=19;}
    void addAtHead(String s)
    {
        if(top<bottom-1)
            arr[top++]=s;
        else
            cout<<"\nDeque is Full";
        }
        void addAtTail(String s)
        {
        if(bottom-1>top)
            arr[bottom-]=s;
        else
            cout<<"\nDeque is Full";
        }
    void forEach();
```

```
void firstThat(char*);
void getLeft();
void getRight();
};
    void Deque::forEach()
    {
        cout<<"\n Top Strings"<<endl;
        for(int i=0;i<top;i++)
            arr[i].show();
        cout<<"\n Bottom Strings"<<endl;
        for(i=19;i>bottom;i-)
            arr[i].show();
    }
    void Deque::firstThat(char* s)
    {
        int count=0;
        for(int i=0;i<20;i++)
        if(arr[i].same(s)==0)count++;
        if(count>0)cout<<"\nStrint matched "<<count<<" times";
        else cout<<"\n String is not found in Deque";
    }
    void Deque::getLeft()
    {
        cout<<"\n Left end items:"<<endl;
        while(top>=0)arr[top-].show();
        cout<<"\n Right end items:"<<endl;
        int b=19;
        while(b>=bottom)arr[b-].show();
    }
    void Deque::getRight()
    {
        cout<<"\n Left end items:"<<endl;
        int t=0;
        while(t<=top)arr[t++].show();
        cout<<"\n Right end items:"<<endl;
        while(bottom<=19)arr[bottom++].show();
    }

int main()
```

```
{
      clrscr();
      String s1("amit");
      String s2("kumar");
      String s3("Verma");
      String s4("Lalit");
      String s5("Kishore");
      String s6("Arora");

      Deque D;
      D.addAtHead(s1);
      D.addAtHead(s2);
      D.addAtHead(s3);

      D.addAtTail(s6);
      D.addAtTail(s5);
      D.addAtTail(s4);

      D.forEach();
      char str[10];
      cout<<"\n Enter String to check:";
      cin>>str;
      D.firstThat(str);
      D.getLeft(); // use only one getLeft() or getRight()
      //D.getRight();
      getch();
      return 0;
}
```

Q.14. Create a base class called shape. Use this class to store two double type values that could be used to compute the area of figures. Derive two specific classes called triangle and rectangle from the base shape. Add to the base class, a member function get_data ( ) to initialise base class data members and another member function display_area ( ) to compute and display the area of figures. Make display_area ( ) as a virtual function and re-define this function in the derived classes to suit their requirements.

Using these three classes, design a program that will accept dimensions of a triangle or a rectangle interactively and display the area.

Remember the two values given as input will be treated as lengths of two sides in the case of rectangles and as base and height in the case of triangles and used as follows:

Area of rectangle = x * y

Area of triangle = ½ * x * y

```
#include<iostream.h>
#include<conio.h>
#include<string.h>
class shape
{
protected:
double x,y;
public:
    void get_data()
    {
        cout<<"\n Enter first side: ";
        cin>>x;
        cout<<"\n Enter Second Side: ";
        cin>>y;
    }
virtual void display_area()=0;
};
class rectangle:public shape
{
public:
void display_area()
{
    cout<<"\n Area of Rectangle : "<<x*y;
}
};
class triangle:public shape
{
public:
    void display_area()
    {
        cout<<"\n Area of Triangle : "<<(x*y)/2.0;
    }
};

int main()
{
    shape *x;
    rectangle R;
    triangle T;
```

```
    clrscr();
    R.get_data();
    x=&R;
    x->display_area();
    T.get_data();
    x=&T;
    x->display_area();
    return 0;
}
```

❑❑❑

APPENDICES

Appendix A

PREDICT THE OUTPUT OR ERROR

Note : All the C sample programs are tested under Turbo C/C++ compilers. It is assumed that,

- Programs run under DOS environment,
- The underlying machine is an x86 system,
- Program is compiled using Turbo C/C++ compiler.

The program output may depend on the information based on this assumptions (for example sizeof(int) == 2 may be assumed).

Following are some C sample questions.

1. Predict the output or error(s) for the following:

```
class Sample
{
public:
    int *ptr;
    Sample(int i)
    {
    ptr = new int(i);
    }
    ~Sample()
    {
    delete ptr;
    }
    void PrintVal()
    {
    cout « "The value is " « *ptr;
    }
};
void SomeFunc(Sample x)
{
cout « "Say i am in someFunc " « endl;
}
int main()
{
```

```
Sample s1= 10;
SomeFunc(s1);
s1.PrintVal();
}
```

Answer:

Say i am in someFunc

Null pointer assignment(Run-time error)

Explanation:

As the object is passed by value to SomeFunc the destructor of the object is called when the control returns from the function. So when PrintVal is called it meets up with ptr that has been freed.The solution is to pass the Sample object by reference to SomeFunc:

```
void SomeFunc(Sample &x)
{
cout « "Say i am in someFunc " « endl;
}
```

because when we pass objects by refernece that object is not destroyed. while returning from the function.

2. Predict the output or error(s) for the following:

```
class base
{
public:
int bval;
base(){ bval=0;}
};

class deri:public base
{
public:
int dval;
deri(){ dval=1;}
};
void SomeFunc(base *arr,int size)
{
for(int i=0; i‹size; i++,arr++)
   cout«arr-›bval;
cout«endl;
}

int main()
{
base BaseArr[5];
```

```
SomeFunc(BaseArr,5);
deri DeriArr[5];
SomeFunc(DeriArr,5);
}
```

Answer:

00000
01010

Explanation:

The function SomeFunc expects two arguments.The first one is a pointer to an array of base class objects and the second one is the sizeof the array.The first call of someFunc calls it with an array of bae objects, so it works correctly and prints the bval of all the objects. When Somefunc is called the second time the argument passed is the pointer to an array of derived class objects and not the array of base class objects. But that is what the function expects to be sent. So the derived class pointer is promoted to base class pointer and the address is sent to the function. SomeFunc() knows nothing about this and just treats the pointer as an array of base class objects. So when arr++ is met, the size of base class object is taken into consideration and is incremented by sizeof(int) bytes for bval (the deri class objects have bval and dval as members and so is of size >= sizeof(int)+sizeof(int)).

3. Predict the output or error(s) for the following:

```
class base
 {
 public:
   void baseFun(){ cout«"from base"«endl;}
 };
 class deri:public base
 {
 public:
   void baseFun(){ cout« "from derived"«endl;}
 };
void SomeFunc(base *baseObj)
{
 baseObj->baseFun();
}
int main()
{
base baseObject;
SomeFunc(&baseObject);
deri deriObject;
SomeFunc(&deriObject);
}
```

Answer:

from base
from base

Explanation:

As we have seen in the previous case, SomeFunc expects a pointer to a base class. Since a pointer to a derived class object is passed, it treats the argument only as a base class pointer and the corresponding base function is called.

```
class base
 {
 public:
   virtual void baseFun(){ cout«"from base"«endl;}
 };
 class deri:public base
   {
 public:
   void baseFun(){ cout« "from derived"«endl;}
   };
void SomeFunc(base *baseObj)
{
   baseObj->baseFun();
}
int main()
{
base baseObject;
SomeFunc(&baseObject);
deri deriObject;
SomeFunc(&deriObject);
}
```

Answer:

from base
from derived

Explanation:

Remember that baseFunc is a virtual function. That means that it supports run-time polymorphism. So the function corresponding to the derived class object is called.

4. Predict the output or error(s) for the following:

```
void main()
{
   int a, *pa, &ra;
   pa = &a;
   ra = a;
   cout «"a="«a «"*pa="«*pa «"ra"«ra ;
}
```

Answer:

Compiler Error: 'ra',reference must be initialised

Explanation:

Pointers are different from references. One of the main differences is that the pointers can be both initialised and assigned, whereas references can only be initialised. So this code issues an error.

5. Predict the output or error(s) for the following:

```
const int size = 5;
void print(int *ptr)
{
   cout«ptr[0];
}

void print(int ptr[size])
{
   cout«ptr[0];
}

void main()
{
   int a[size] = {1,2,3,4,5};
   int *b = new int(size);
   print(a);
   print(b);
}
```

Answer:

Compiler Error : function 'void print(int *)' already has a body

Explanation:

Arrays cannot be passed to functions, only pointers (for arrays, base addresses) can be passed. So the arguments int *ptr and int prt[size] have no difference as function arguments. In other words, both the functoins have the same signature and so cannot be overloaded.

6. Predict the output or error(s) for the following:

```
class some{
public:
       ~some()
       {
           cout«"some's destructor"«endl;
       }
   };

void main()
{
       some s;
       s.~some();
```

```
}
```

Answer:

some's destructor some's destructor

Explanation:

Destructors can be called explicitly. Here 's.~some()' explicitly calls the destructor of 's'. When main() returns, destructor of s is called again, hence the result.

7. Predict the output or error(s) for the following:

```
#include <iostream.h>
class fig2d
{
    int dim1;
    int dim2;
public:
    fig2d() { dim1=5; dim2=6;}
    virtual void operator«(ostream & rhs);
};
void fig2d::operator«(ostream &rhs)
{
    rhs «this->dim1«" "«this->dim2«" ";
}

/*class fig3d : public fig2d
{
    int dim3;
public:
    fig3d() { dim3=7;}
    virtual void operator«(ostream &rhs);
};
void fig3d::operator«(ostream &rhs)
{
    fig2d::operator «(rhs);
    rhs«this -> dim3;
}
*/

void main()
{
    fig2d obj1;
// fig3d obj2;

    obj1 « cout;
// obj2 « cout;
}
```

Answer:

```
5 6
```

Explanation:

In this program, the « operator is overloaded with ostream as argument. This enables the 'cout' to be present at the right-hand-side. Normally, 'cout' is implemented as global function, but it doesn't mean that 'cout' is not possible to be overloaded as member function.

Overloading « as virtual member function becomes handy when the class in which it is overloaded is inherited, and this becomes available to be overrided. This is as opposed to global friend functions, where friend's are not inherited.

8. Predict the output or error(s) for the following:

```
class opOverload{
public:
   bool operator==(opOverload temp);
};

bool opOverload::operator==(opOverload temp){
   if(*this == temp ){
       cout«"The both are same objects\n";
       return true;
   }
   else{
       cout«"The both are different\n";
       return false;
   }
}

void main(){
   opOverload a1, a2;
   a1= =a2;
}
```

Answer:

Runtime Error: Stack Overflow

Explanation:

Just like normal functions, operator functions can be called recursively. This program just illustrates that point, by calling the operator == function recursively, leading to an infinite loop.

9. Predict the output or error(s) for the following:

```
class complex{
   double re;
   double im;
public:
   complex() : re(1),im(0.5) {}
```

```
    bool operator==(complex &rhs);
    operator int(){}
};

bool complex::operator == (complex &rhs){
    if((this->re == rhs.re) && (this->im == rhs.im))
        return true;
    else
        return false;
}

int main(){
    complex c1;
    cout« c1;
}
```

Answer:

Garbage value

Explanation:

The programmer wishes to print the complex object using output re-direction operator,which he has not defined for his lass.But the compiler instead of giving an error sees the conversion function and converts the user defined object to standard object and prints some garbage value.

10. Predict the output or error(s) for the following:

```
class complex{
    double re;
    double im;
public:
    complex() : re(0),im(0) {}
    complex(double n) { re=n,im=n;};
    complex(int m,int n) { re=m,im=n;}
    void print() { cout«re; cout«im;}
};

void main(){
    complex c3;
    double i=5;
    c3 = i;
    c3.print();
}
```

Answer:

5,5

Explanation:

Though no operator= function taking complex, double is defined, the double on the rhs is converted into a temporary object using the single argument constructor taking double and assigned to the lvalue.

11. Predict the output or error(s) for the following:

```
void main()
{
    int a, *pa, &ra;
    pa = &a;
    ra = a;
    cout «"a="«a «"*pa="«*pa «"ra"«ra ;
}
```

Answer:

Compiler Error: 'ra',reference must be initialized

Explanation:

Pointers are different from references. One of the main differences is that the pointers can be both initialised and assigned, whereas references can only be initialized. So this code issues an error.

❑❑❑

Appendix **B**

INTERVIEW RELATED QUESTIONS

Q.1. What is a modifier?

Ans. A modifier, also called a modifying function is a member function that changes the value of at least one data member. In other words, an operation that modifies the state of an object. Modifiers are also known as 'mutators'.

Q.2. What is an accessor?

Ans. An accessor is a class operation that does not modify the state of an object. The accessor functions need to be declared as const operations.

Q.3. Differentiate between a template class and class template.

Ans.Template class: A generic definition or a parameterised class not instantiated until the client provides the needed information. It's jargon for plain templates.

Class template: A class template specifies how individual classes can be constructed much like the way a class specifies how individual objects can be constructed. It's jargon for plain classes.

Q.4. When does a name clash occur?

Ans. A name clash occurs when a name is defined in more than one place. For example, two different class libraries could give two different classes the same name. If you try to use many class libraries at the same time, there is a fair chance that you will be unable to compile or link the program because of name clashes.

Q.5. Define namespace.

Ans. It is a feature in c++ to minimise name collisions in the global namespace. This namespace keyword assigns a distinct name to a library that allows other libraries to use the same identifier names without creating any name collisions. Furthermore, the compiler uses the namespace signature for differentiating the definitions.

Q.6. What is the use of 'using' declaration.

Ans. A using declaration makes it possible to use a name from a namespace without the scope operator.

Q.7. What is an Iterator class?

Ans. A class that is used to traverse through the objects maintained by a container class. There are five categories of iterators:

1. input iterators,
2. output iterators,

3. forward iterators,
4. bidirectional iterators,
5. random access.

An iterator is an entity that gives access to the contents of a container object without violating encapsulation constraints. Access to the contents is granted on a one-at-a-time basis in order. The order can be storage order (as in lists and queues) or some arbitrary order (as in array indices) or according to some ordering relation (as in an ordered binary tree). The iterator is a construct, which provides an interface that, when called, yields either the next element in the container, or some value denoting the fact that there are no more elements to examine. Iterators hide the details of access to and update of the elements of a container class. The simplest and safest iterators are those that permit read-only access to the contents of a container class. The following code fragment shows how an iterator might appear in code:

```
cont_iter:=new cont_iterator();
 x:=cont_iter.next();
 while x/=none do
 ...
 s(x);
 ...
 x:=cont_iter.next();
 end;
```

In this example, cont_iter is the name of the iterator. It is created on the first line by instantiation of cont_iterator class, an iterator class defined to iterate over some container class, cont. Succesive elements from the container are carried to x. The loop terminates when x is bound to some empty value. (Here, none)In the middle of thc loop, there is s(x) an operation on x, the current element from the container. The next element of the container is obtained at the bottom of the loop.

Q.8. List out some of the OODBMS available.

Ans.
- GEMSTONE/OPAL of Gemstone systems.
- ONTOS of Ontos.
- Objectivity of Objectivity inc.
- Versant of Versant object technology.
- Object store of Object Design.
- ARDENT of ARDENT software.
- POET of POET software.

Q.9. List out some of the object oriented methodologies.

Ans.
- Object Oriented Development (OOD) (Booch 1991,1994).
- Object Oriented Analysis and Design (OOA/D) (Coad and Yourdon 1991).
- Object Modelling Techniques (OMT) (Rumbaugh 1991).
- Object Oriented Software Engineering (Objectory) (Jacobson 1992).
- Object Oriented Analysis (OOA) (Shlaer and Mellor 1992).
- The Fusion Method (Coleman 1991).

Q.10. What is an incomplete type?

Ans. Incomplete types refers to pointers in which there is non availability of the implementation of the referenced location or it points to some location whose value is not available for modification.

Example:

int *i=0x400 // i points to address 400 *i=0; //set the value of memory location pointed by i.

Incomplete types are otherwise called uninitialised pointers.

Q.11. What is a dangling pointer?

Ans. A dangling pointer arises when you use the address of an object after its lifetime is over. This may occur in situations like returning addresses of the automatic variables from a function or using the address of the memory block after it is freed.

Q.12. Differentiate between the message and method.

Ans.

Message	Method
Objects communicate by sending messages.	Provides response to a message to each other.
A message is sent to invoke a method.	It is an implementation of an operation.

Q.13. What is an adaptor class or Wrapper class?

Ans. A class that has no functionality of its own. Its member functions hide the use of a third party software component or an object with the non-compatible interface or a non- object oriented implementation.

Q.14. What is a Null object?

Ans. It is an object of some class whose purpose is to indicate that a real object of that class does not exist. One common use for a null object is a return value from a member function that is supposed to return an object with some specified properties but cannot find such an object.

Q.15. What is class invariant?

Ans. A class invariant is a condition that defines all valid states for an object. It is a logical condition to ensure the correct working of a class. Class invariants must hold when an object is created, and they must be preserved under all operations of the class. In particular all class invariants are both preconditions and post-conditions for all operations or member functions of the class.

Q.16. What do you mean by Stack unwinding?

Ans. It is a process during exception handling when the destructor is called for all local objects between the place where the exception was thrown and where it is caught.

Q.17. Define precondition and post-condition to a member function.

Ans. Precondition: A pre-condition is a condition that must be true on entry to a member function. A class is used correctly if preconditions are never false. An operation is not responsible for doing anything sensible if its pre-condition fails to hold. For example, the interface invariants of stack class say nothing about pushing yet another element on a stack that is already full. We say that isful() is a pre-condition of the push operation.

Post-condition: A post-condition is a condition that must be true on exit from a member function if the pre-condition was valid on entry to that function. A class is implemented correctly if post-conditions are never false. >For example, after pushing an element on the stack, we know that isempty() must necessarily hold. This is a post-condition of the push operation.

Q.18. What are the conditions that have to be met for a condition to be an invariant of the class?

Ans. The condition should hold at the end of every constructor. The condition should hold at the end of every mutator(non-const) operation.

Q.19. What are proxy objects?

Ans. Objects that stand for other objects are called proxy objects or surrogates.

Example:

```
Template <class T>
    class Array2D
    {
        public:
            class Array1D
            {
                public:
            T& operator[] (int index);
            const T& operator[] (int index) const;
            ...
            };
            Array1D operator[] (int index);
            const Array1D operator[] (int index) const;
            ...
        };
```

The following then becomes legal:

```
Array2Ddata(10,20);
........
cou<<data[3][6]; // fine
```

Here data[3] yields an Array1D object and the operator [] invocation on that object yields the float in position(3,6) of the original two-dimensional array. Clients of the Array2D class need not be aware of the presence of the Array1D class. Objects of this latter class stand for one-dimensional array objects that, conceptually, do not exist for clients of Array2D. Such clients program as if they were using real, live, two-dimensional arrays. Each Array1D object stands for a one-dimensional array that is absent from a conceptual model used by the clients of Array2D. In the above example, Array1D is a proxy class. Its instances stand for one-dimensional arrays that, conceptually, do not exist.

Q.20. Name some pure object oriented languages.

Ans.
- Smalltalk,
- Java,
- Eiffel,
- Sather.

Q.21. Name the operators that cannot be overloaded.

Ans. sizeof . .* .-> :: ?:

Q.22. What is a node class?

Ans. A node class is a class that,

- relies on the base class for services and implementation,
- provides a wider interface to te users than its base class,
- relies primarily on virtual functions in its public interface
- depends on all its direct and indirect base class
- can be understood only in the context of the base class
- can be used as base for further derivation
- can be used to create objects.
- A node class is a class that has added new services or functionality beyond the services inherited from its base class.

Q.23. What is an orthogonal base class?

Ans. If two base classes have no overlapping methods or data they are said to be independent of, or orthogonal to each other. Orthogonal in the sense means that two classes operate in different dimensions and do not interfere with each other in any way. The same derived class may inherit such classes with no difficulty.

Q.24. What is a container class? What are the types of container classes?

Ans. A container class is a class that is used to hold objects in memory or external storage. A container class acts as a generic holder. A container class has a predefined behavior and a well-known interface. A container class is a supporting class whose purpose is to hide the topology used for maintaining the list of objects in memory. When a container class contains a group of mixed objects, the container is called a heterogeneous container; when the container is holding a group of objects that are all the same, the container is called a homogeneous container.

Q.25. What is a protocol class?

Ans. An abstract class is a protocol class if:

- it neither contains nor inherits from classes that contain member data, non-virtual functions, or private (or protected) members of any kind.
- it has a non-inline virtual destructor defined with an empty implementation,
- all member functions other than the destructor including inherited functions, are declared pure virtual functions and left undefined.

Q.26. What is a mixin class?

Ans. A class that provides some but not all of the implementation for a virtual base class is often called mixin. Derivation done just for the purpose of redefining the virtual functions in the base classes is often called mixin inheritance. Mixin classes typically don't share common bases.

Q.27. What is a concrete class?

Ans. A concrete class is used to define a useful object that can be instantiated as an automatic variable on the program stack. The implementation of a concrete class is defined. The concrete class is not intended to be a base class and no attempt to minimise dependency on other classes in the implementation or behaviour of the class.

Q.28. What is the handle class?

Ans. A handle is a class that maintains a pointer to an object that is programmatically accessible through the public interface of the handle class.

Explanation:

In case of abstract classes, unless one manipulates the objects of these classes through pointers and references, the benefits of the virtual functions are lost. User code may become dependent on details of implementation classes because an abstract type cannot be allocated statistically or on the stack without its size being known. Using pointers or references implies that the burden of memory management falls on the user. Another limitation of abstract class object is of fixed size. Classes however are used to represent concepts that require varying amounts of storage to implement them. A popular technique for dealing with these issues is to separate what is used as a single object in two parts: a handle providing the user interface and a representation holding all or most of the object's state. The connection between the handle and the representation is typically a pointer in the handle. Often, handles have a bit more data than the simple representation pointer, but not much more. Hence the layout of the handle is typically stable, even when the representation changes and also that handles are small enough to move around relatively freely so that the user needn't use the pointers and the references.

Q.29. What is a parameterised type?

Ans. A template is a parameterised construct or type containing generic code that can use or manipulate any type. It is called parameterized because an actual type is a parameter of the code body. Polymorphism may be achieved through parameterized types. This type of polymorphism is called parameteric polymorphism. Parameteric polymorphism is the mechanism by which the same code is used on different types passed as parameters.

Q.30. What is a smart pointer?

Ans. A smart pointer is an object that acts, looks and feels like a normal pointer but offers more functionality. In C++, smart pointers are implemented as template classes that encapsulate a pointer and override standard pointer operators. They have a number of advantages over regular pointers. They are guaranteed to be initialised as either null pointers or pointers to a heap object. Indirection through a null pointer is checked. No delete is ever necessary. Objects are automatically freed when the last pointer to them has gone away. One significant problem with these smart pointers is that unlike regular pointers, they don't respect inheritance. Smart pointers are unattractive for polymorphic code. Given below is an example for the implementation of smart pointers.

Example:

```
template <class X>
class smart_pointer
{
    public:
        smart_pointer(); // makes a null pointer
        smart_pointer(const X& x) // makes pointer to copy of x

        X& operator *( );
        const X& operator*( ) const;
```

```
            X* operator->() const;
            smart_pointer(const smart_pointer <X> &);
            const smart_pointer <X> & operator
            =(const smart_pointer<X>&);
            ~smart_pointer();
        private:
            //...
    };
```

This class implement a smart pointer to an object of type X. The object itself is located on the heap. Here is how to use it: smart_pointer ‹employee› p= employee("Harris",1333); Like other overloaded operators, p will behave like a regular pointer, cout<<*p; p -> raise_salary(0.5);

Q.31. What is reflexive association?

Ans. The 'is-a' is called a reflexive association because the reflexive association permits classes to bear the is-a association not only with their super-classes but also with themselves. It differs from a 'specializes-from' as 'specialises-from' is usually used to describe the association between a super-class and a sub-class. For example, Printer is-a printer.

Q.32. What is slicing?

Ans. Slicing means that the data added by a sub-class are discarded when an object of the sub-class is passed or returned by value or from a function expecting a base class object.

Explanation: Consider the following class declaration:

```
    class base
    {
    ...
    base& operator =(const base&);
    base (const base&);
}
void fun( )
{
    base e=m;
    e=m;
}
```

As base copy functions don't know anything about the derived only the base part of the derived is copied. This is commonly referred to as slicing. One reason to pass objects of classes in a hierarchy is to avoid slicing. Other reasons are to preserve polymorphic behaviour and to gain efficiency.

Q.33. What are proxy objects?

Ans. Objects that points to other objects are called proxy objects or surrogates. Its an object that provides the same interface as its server object but does not have any functionality. During a method invocation, it routes data to the true server object and sends back the return value to the object.

Q.34. Differentiate between declaration and definition in C++.

Ans. A declaration introduces a name into the program; a definition provides a unique description of an entity (*e.g.,* type, instance, and function). Declarations can be repeated in a given scope, it introduces a name in a given scope. There must be exactly one definition of every object, function or class used in a C++ program. A declaration is a definition unless:

- it declares a function without specifying its body,
- it contains an extern specifier and no initialiser or function body,
- it is the declaration of a static class data member without a class definition,
- it is a class name definition,
- it is a typedef declaration.

A definition is a declaration unless:

- it defines a static class data member,
- it defines a non-inline member function.

Q.35. Describe the main characteristics of static functions.

Ans. The main characteristics of static functions include:

- It is without the a this pointer,
- It can't directly access the non-static members of its class
- It can't be declared const, volatile or virtual.
- It doesn't need to be invoked through an object of its class, although for convenience, it may.

Q.36. Will the inline function be compiled as the inline function always? Justify.

Ans. An inline function is a request and not a command. Hence it won't be compiled as an inline function always.

Explanation: Inline-expansion could fail if the inline function contains loops, the address of an inline function is used, or an inline function is called in a complex expression. The rules for inlining are compiler dependent.

Q.37. Define a way other than using the keyword inline to make a function inline.

Ans. The function must be defined inside the class.

Q.38. How can a '::' operator be used as unary operator?

Ans. The scope operator can be used to refer to members of the global namespace. Because the global namespace doesnâ•™t have a name, the notation :: member-name refers to a member of the global namespace. This can be useful for referring to members of global namespace whose names have been hidden by names declared in nested local scope. Unless we specify to the compiler in which namespace to search for a declaration, the compiler simple searches the current scope, and any scopes in which the current scope is nested, to find the declaration for the name.

❑❑❑

Appendix **C**

C++ TIPS

This is not a code guideline document. This is more of a document to fuel questions while you design and code with C++. In some cases the point is simply stated and probably comes off as a rule. In reality, they are simply meant as rules of thumb.

One of the main problems in C++ (more so than C) is that C++ provides many mechanisms in the language by which the same task can be achieved through different policies. For example, C++ has the Polymorphism mechanism. Some of the policies are templates, macros, inheritance, overloading,

(those are the static, compile-time ones), and virtual functions (the run-time polymorphic policy). Hopefully the following will provide enough fuel for questions to arrive at the best policy to use for a particular design.

I. INHERITANCE

1. Abstract Base Class (ABC) : Make ABC class constructor protected when possible. Derived classes can have public constructor to override this. The same is true for non-ABC classes as well.

2. Class Inheritance : Use protected keyword where ever possible, never use public to expose data members so the inheriting classes have access to them.

3. Multiple Inheritance : The only drawback is the most derived class must initialise the lowest base classes. This breaks encapsulation. Most people view multiple inheritance as a bad thing, but when used sparingly for parts of a design, there is no problem with it.

4. Data in Classes should be keep Private as much as Possible: Use a member function if a client needs to access the data. If a member function does not need to be seen by a client, make it private. If the class might be inherited, then protected may be a good choice.

5. Private Inheritance: Don't use it. Too many ambiguities when used with runtime type identification. Use a private data member instead or use public inheritance.

Example:class Foo: private Bar { ... } dont do

6. Inheritance & Virtual Overrides: Care must be taken in overriding inherited functions. Sometimes functions are grouped together, and all need to be overridden. The base class designer must make this clear, if overriding one function requires multiple functions to be overridden.

7. Inheritance & Get/Set functions: Typically functions that perform Get/Set shouldn't be overridden unless they are used by the derived class. Otherwise if the base class does direct field manipulation, you usually can't override it correctly, or it could be a maintenance nightmare.

Note: Get/Set functions as referred to above are merely meant as abstractions. As a rule of thumb (thumb must be getting pretty big by now), one should not create functions called Get or Set, or flavors thereof. They tend to break the spirit of encapsulation. Of course, there will be times to use them.

II. SCOPE

1. Law of Demeter : Do not make references to classes in a class that are not variables of the class, inherited by the class, or global. This also applies to including header files.

Example:

```
class Foo{public: Go(){} };
class Bar{ Foo aFoo; public:   Foo GetFoo(){return aFoo;} };
class Fubar{ public: void Bad(){ Bar aBar; aBar.GetFoo().Go();} };
```

The method Bad() breaks encapsulation. It calls a method of a class it probably does not need to know about. This will also affect maintainability. If the Foo class changes, the changes may also need to be done in the Fubar class.

The other side of the coin that must be looked at is do you want a pass-through method in the Bar class that simply calls the Go() method of the Foo class. Lots of silly simple 1 line member functions may not be desirable in all the classes.

What probably needs to be looked at to decide what road to take is speed, or perhaps redesign the classes. Beyond the Law of Demeter is Doug's rule of thumb: Don't play hide and seek with data.

2. Scope: Another way to say the previous point is to keep scope small. This will increase the lifetime of the code and keep it maintainable and safe.

III. CLASSES

1. Be Explicit about the keywords, Public, Protected, and Private in a Class Interface. Try not to have multiple sections. In other words, multiple private sections in the class interface. It is generally a good idea to place the public section first because this is what most people are looking for when they go to use a class.

2. Make Classes as Atomic as Possible. Give them a clear purpose. Break them into smaller classes if they become too complex. This may also eliminate duplicate code.

3. Don't let the Compiler Generate the Default Constructor, Destructor, Operator= and Copy Constructor. If the class is entirely value based, this is probably fine, if not, for example, if the class has a data member that is a pointer, the above will probably not work. Note, the default copy constructor only does a memcpy of the class, so all you copy is the pointer data. This may not be sufficient for copying a class. Regardless, if you do want the default ones, place them in the code, and leave them commented.

For example:

```
// Fubar(const Fubar&) use default copy constructor
```

4. If your class contains pointers, you should create the default constructor destructor, operator=, and copy constructor.

5. Class Copy: If the class should never be copied, then place the copy constructor in the private section and don't implement it. The linker will catch this, and the program will fail to build. This, although not graceful, is better than a malformed program.

6. Initialisation: Perform all data member initialisation in the constructor. It's best not to leave uninitialised objects running around in the system. Note, it is often more efficient using the constructor initialiser list, otherwise, the default constructor would be called, and then you probably call member functions of the object later in the constructor. For example:

```
class Foo{ Bar mung; Foo(int iCount) : mung(iCount) {} ... };
```

The variable mung is initialised once. But in the following:

```
class Foo{ Bar mung; Foo(int iCount){ mung = iCount; } ... };
```

mung's default constructor is called before the body of the class Foo's constructor is entered. Then mung is set again - this assumes that mung has an assignment operator. The net effect is that mung is initialised twice.

7. Class Naming: There exists several ways to name classes that seem to work well for certain groups or people.

IV. MISC

1. Implicit int: The 'implicit int' rule will go away in the next C++ standard. So for a proto like: 'main()' you will have to say 'int main()' in the future. Same for variables.

2. Pre-processor: Avoid it. Use const values in the class, or inline functions instead of macros. This is not to say, never use any #defines.

V. OVERLOADING

1. Operator Overloading: It's syntactic sugar. Don't add them if they are not needed. This does NOT refer to the typical ones like '=', '==', but ones like '()', '[]', '+'. It does not always make sense to add two objects together.

2. Overloading: If a member function is conditionally executing code, it may be a candidate for operator overloading, or just overloading.

3. Operator Overloading and Chaining: When designing an overloaded operator, think about whether it needs to be chained. For example:

```
String cstr = "a" + "b" + "c";
```

The String class's operator returns a reference to the String class. A partial implementation might be:

```
class String
{
public:
    String& operator +(const char *pcBuf)
    {
     // code to add the char* to the string
     return *this;
    }
    ...
};
```

VI. PARAMETERS

1. Argument Passing: The first choice is typically a const ref. The const ref is basically an alias, and is easier to use than a pointer. It creates the same amount of instruction code as passing a pointer (for most cases). It's typically better than passing by value, where an object constructor will be called (if its an object). As a rule of thumb, you might want too give the following a whirl:

IN	const &	
OUT	&	If the object has the support functs
INOUT	*&	Acts like a **.

So for an IN parameter, what the 'const &' says, is here is a reference to it, but you cannot modify the object. But you can call member functions that do not change the object (member functions defined as const). The OUT parameter is a parameter that is passed to a function that will modify it. If the parameter needs to be created, then the INOUT parameter of *& may be a good choice.

2. Returning Ref: In functions that return a reference, remember not to reference a temporary object and return it. For example:

String &Zippo(void){ .... return String();

What happens, is that the String() is a temporary object that upon return goes out of scope and is destroyed. Thus, you return a reference to a destroyed object. Unfortunately, the program will probably work in most cases till it's shipped.

3. Ref vs Pointer: Here's another way to look at when to use references, and when should to use pointers.

C programmers sometimes don't like references since the reference semantics they provide isn't *explicit* in the caller's code. After a bit of C++ experience, however, one quickly realises this "information hiding" is an asset rather than a liability. In particular, reuse-centered OOP tends to migrate the level of abstraction away from the language of the machine toward the language of the problem. References are usually preferred over ptrs whenever you don't need "re-seating". This usually means that references are most useful in a class' public interface. References then typically appear on the skin of an object, and pointers on the inside. The exception to the above is where a function's parameter or return value needs a "sentinel" reference. This is usually best done by returning/taking a pointer, and giving the nil ptr (0) this special significance (references should always alias *objects*, not a dereferenced nil ptr).

VII. CONSTRUCTORS (more on classes)

1. Creating Constructors: These creatures should be simple. Try not to do anything in them that may generate errors. Remember they don't have return values. If it's necessary to allocate memory or other complex things in the constructor, throw an exception if possible as the first recourse. Else, the class will have to beprotected everywhere that may use something that may be in error. It's generally not a pretty sight to see an error returned in the constructor's signature.

2. Constructors: Another reason to keep them simple is in the case of inheritance.

3. Destructors: It's responsibility is to release resources allocated during the class's lifetime, not just from construction.

4. Member Initialisation List: The constructor can have a comma separated member initialisation list. If a class contains value based classes as data members, they can be initialised in the constructor.

For example:

```
class Foo{
    String cstr1;     // value based class called String
    String cstr2;     // value based class called String
public:
    Foo(const char* pcStr1, const char*pcStr2):
        cstr1(pcStr1), cstr2(pcStr2){}
};
```

With the variables cstr1, and cstr2 initialized in the constructor, they are initialized only once. Else, if they were initialized in the body of the constructor, they would first be initialized with a default constructor, then again in the body.

VIII. EXCEPTIONS

1. Exceptions: Use exception hierarchies, possibly even derived from the Standard C++ ones.

2. Exceptions: Throw exceptions by value and catch them by reference. This way the exception handling mechanism cleans up anything created on the heap. If you throw exceptions by pointer, the catcher must know how to destroy them. This is probably not a good coupling. Even so, any up casting may slice and dice the object.

IX. TEMPLATES (more on classes)

1. Templates: Before creating new ones, see if they are part of the C++ Standard.

2. Templates: When creating them, try to filter out any code that does not depend on the type, and place that into a base class.Thus, the template class itself is only the necessary information that depends on type.

INDEX

P

R

S

T

V

W

❑❑❑

Win Prizes

ATTENTION READERS

Win Prizes

We request for your frank assessment regarding various aspects of the book as given below :

OBJECT ORIENTED SYSTEMS WITH C++

By: Lalit Kishore Arora

Please Fill up the Sheet in Neat and Clean Writing

(i) What topics of your syllabus, important for your examination point of view, are not covered in this book ?

....

....

....

....

(ii) What are the chapters wherein the treatment of the subject matter is not systematic or organised or updated ?

....

....

....

....

(iii) Have you come across misprints/mistakes/factual inaccuracies in the book? Please specify the chapters and the page numbers.

....

....

....

....

(iv) What other books on the same subject have you found/heard better than the present book? Please specify in terms of price and quality

....

....

....

....

(v) Further suggestions and comments for the improvement of this book.

....

....

....

....

(PS: If need be please attach a separate sheet to write your views)

OTHER DETAILS

(i) Who recommended you this book ? (Please tick (✓) in the box near the option relevant to you.

☐ Teacher ☐ Friends ☐ Bookseller

....

....

....

....

(ii) Name of the recommending teacher, his designation and address :

....

....

....

....

(iii) Name and address of the bookseller you purchased the book from :

....

....

....

....

(iv) Name and address of your institution (Please mention the University or Board as the case may be)

....

....

....

....

(v) Your name and complete postal address :

....

....

....

....

(vi) Course you are enrolled in :

....

....

....

....

The best assessment will be awarded each month: The award will be in the form of our publications as decided by the Editorial Board.

Please mail the filled up coupon at your earliest to :

S.K. KATARIA & SONS®

4885/109, Prakash Mahal, Dr. Subhash Bhargav Lane,
Opposite Delhi Medical Association, Daryaganj, New Delhi–110002 (INDIA)
Phone: +91-11-43551243; Mobile: +91-9871775858, 7678425209
e-mail: katariabook@gmail.com, katariabooks@yahoo.com
Website: www.skkatariaandsons.com